DONALD D. VOISINET

Professor of Technology
Niagara County Community College
Sanborn, New York

Residential and Commercial Electrical Design Projects

PRENTICE-HALL, INC., Englewood Cliffs, New Jersey 07632

Library of Congress Cataloging in Publication Data

Voisinet, Donald D., N.D.
  Residential and commercial electrical design
projects.

  Includes index.
  1. Electric wiring, Interior.   I. Title.
  TK3271.V65 1985   621.31'924   84-16030
  ISBN 0-13-774688-1
  ISBN 0-13-774670-9 (pbk.)

Editorial/production supervision
     and interior design: Inkwell
Cover design: Whitman Studios, Inc.
Manufacturing buyer: Gordon Osbourne

National Electrical Code® is a registered trademark of the National Fire
Protection Association, Inc., Quincy, MA for a triennial electrical
publication. The term, National Electrical Code, as used herein means
the triennial publication constituting the National Electrical Code and
is used with permission of the National Fire Protection Association, Inc.

Printed in the United States of America

10 9 8 7 6 5 4 3 2 1

ISBN   0-13-774688-1

ISBN   0-13-774670-9 {PBK.}

PRENTICE-HALL INTERNATIONAL, INC., *London*
PRENTICE-HALL OF AUSTRALIA PTY. LIMITED, *Sydney*
EDITORA PRENTICE-HALL DO BRASIL, LTDA., *Rio de Janeiro*
PRENTICE-HALL CANADA INC., *Toronto*
PRENTICE-HALL OF INDIA PRIVATE LIMITED, *New Delhi*
PRENTICE-HALL OF JAPAN, INC., *Tokyo*
PRENTICE-HALL OF SOUTHEAST ASIA PTE. LTD., *Singapore*
WHITEHALL BOOKS LIMITED, *Wellington, New Zealand*

Contents

Preface

All residential and commercial building construction requires some type of electrical power distribution. To accomplish this a set of working drawings must be prepared. The working drawings are created by the application of design theory using the appropriate codes. This book provides an in-depth treatment of the entire process. Each chapter begins at an elementary level and advances very rapidly to actual on-the-job applications. As a commercial electrical design drafter, you will encounter numerous special conditions. Consequently, the full spectrum of building construction activity is covered. This provides valuable experience when such situations arise.

Sufficient theory is presented in each chapter so that the projects can be completed. If desired, however, a theory text may accompany this electrical design drafting project book. The problem encountered with the use of only a theory textbook is that it lacks project applications. Gaining a complete background requires substantial application work. The outcome of each project in the text requires a working layout drawing. Each may be used for electrical power and distribution construction.

This book is useful to anyone involved with residential, commercial, and industrial design drafting. This includes engineers, architects, designers and drafters. Before using it, however, you must meet certain prerequisites. You must have a knowledge of drafting or at least fundamentals of engineering drawing. In addition, a minimum level of technique should have been developed so that each drawing may be professionally prepared. It does not matter whether your technique has been developed traditionally or by computer (CAD). In either case the drawings can be generated.

Completion of the projects in this book will give you a distinct advantage during a career search. You will know and be able to do things that others cannot. Good luck and success in your endeavors.

Donald D. Voisinet

1

General Electrical Information

1-1 CODES AND STANDARDS

Commercial and residential construction is governed by various building codes. The electrical segment is considered more dangerous than other segments, such as structural and plumbing. Thus a series of codes and standards has been established. Its purpose is the practical safeguarding of the hazards of using electricity. Hazards often occur due to the overloading of individual circuitry. Several professional organizations have been involved in the creation of code requirements, standards, and recommended practices. The most prominent include:

1. National Fire Protection Association (NFPA)
2. National Electrical Manufacturers Association (NEMA)
3. Underwriters' Laboratories, Inc. (UL)
4. Illuminating Engineering Society (IES)
5. American National Standards Institute, Inc. (ANSI)

The document considered to be the electrical construction "bible" is the *National Electrical Code*® (NEC). This book, shown in Figure 1-1, specifies national electrical requirements.[1] Additionally, states and locales may impose more stringent regulations. The *National Electrical Code*® is under the sponsorship of the National Fire Protection Association and should be consulted whenever performing electrical construction work. Anytime it states "shall," compliance is mandatory. Legal enforcement of the code may be made by the "authority having jurisdiction." This is normally the building inspector. When analyzing a specific situation, this text refers to the appropriate Article. Each Article, however, should be consulted separately since there are many special features, exceptions, and so on. For example, copper conductors will normally be utilized, but aluminum conductors may be used in some cases. It is not allowed to connect

---

[1] *National Electrical Code*® is a registered trademark of the National Fire Protection Association, Inc., Boston, Massachusetts, for a triennial electrical publication. The term *National Electrical Code*® as used herein means the triennial publication constituting the *National Electrical Code*® and is used with permission of the National Fire Protection Association, Inc.

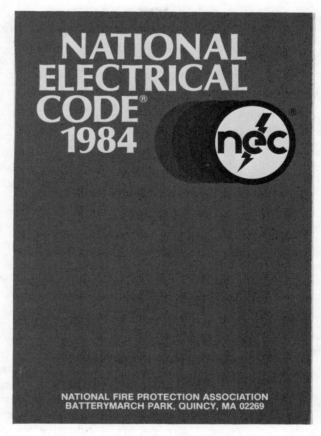

Figure 1-1 NEC book (1984 edition)

these dissimilar metals in a common splicing device, however, due to the occurrence of galvanic action. Another example is the use of appropriate conductor insulation for a particular environmental condition. Certain locations may be dry, whereas others may be corrosive. The correct insulation approved for the specific condition must be used. These examples serve to demonstrate the need to consult the code.

The National Fire Protection Association, referred to as NFPA, holds the NEC copyright. It is located at

470 Atlantic Avenue
Boston, MA 02210

All communication regarding the NEC should be sent to the NFPA headquarters. This includes questions and document referencing. Even though the NFPA must be consulted, it has no power or authority to police or enforce document compliance. The Department of Labor, through the Occupational Safety and Health Administration (OSHA), will police it where OSHA is enforced.

The National Electrical Manufacturers Association, referred to as NEMA, has developed many standards for manufacturer equipment specification. This professional organization is located at

2101 L Street N.W.
Washington, DC 20037

Reference to NEMA is made when specifying equipment such as a motor frame size or a protective enclosure. A NEMA-type enclosure is numerically designated. Each number refers to a different enclosure for a different application. For example, a NEMA type 4 enclosure is moisture proof for interior and exterior use; a NEMA type 12 is dust proof; a NEMA type 7 is explosion proof; and a NEMA type 9 is hazardous dust proof. If the application is not special, a general-purpose NEMA type 1 may be used. This type, as shown in Figure 1-2, prevents hands or tools from accidentally touching live electrical contacts.

The Underwriters' Laboratory, referred to as U.L., is an independent laboratory located at

207 East Ohio Street
Chicago, IL 60611

Any product tested and found safe by the U.L. is authorized to carry the U.L. label.

The Illuminating Engineering Society is referred to as IES and is located at

345 East 47th Street
New York, NY 10017

The IES has compiled and recommended practices concerning illumination. It is the generally recognized standard to use for the design and specifications of lighting and luminaire systems. In the interest of energy conservation, some state organizations have recommended lower illumination levels. Some of these recommendations, however, are counter-productive to good lighting practice. Thus this text will utilize only IES standards.

The American National Standards Institute (ANSI) is located at

1430 Broadway Avenue
New York, NY 10020

It is responsible for setting numerous codes and standards. All drafting standards fall within its domain. The ANSI number for the *National Electrical Code* is ANSI CI-1984.

1-2 SOURCE OF POWER

Electrical energy is required for a power distribution system. Several types of energy sources are found in nature. The most common means to generate electricity is the use of *fossil fuels*. These include natural

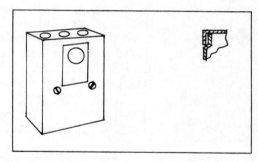

Figure 1-2 NEMA type 1 general-purpose enclosure

gas, oil, and coal. These fuels are burned at power plants similar to the one shown in Figure 1-3(a). The heat generated is converted to electricity. It is then brought to the consumer by means of high-voltage transmission lines. A voltage transformer steps down the voltage to a usable level prior to consumption.

Uranium is also used to generate electricity. The reaction that occurs in a nuclear *fission* reactor as shown in Figure 1-3(b) converts the uranium to plutonium, while giving off large amounts of heat. This method of conversion has inherent problems. Plutonium is highly radioactive and has a half-life decay of more than 24,000 years. This means that the plutonium will still be at approximately half the level of radioactivity in the year 26,000. At present, there is no available method to dispose properly of this waste product. Because of these problems, popular opinion, government regulations, construction, and compliance costs, nuclear fission plant construction has been reduced to nearly zero. Not a single plant has been approved for new construction. Another type of nuclear reaction, *fusion*, would create no plutonium by-product, yet generate enough power to satisfy all of our needs for centuries. Fusion is similar to the sun's reaction. The problem with a fusion reaction is that tremendously high temperatures and pressures are required for it to occur. At present, this cannot be accomplished commercially.

As the availability of fossil fuels decreases, alternative energy sources will become more economically feasible. Besides hydro reservoirs, the following sources have varying degrees of future potential: seawater, tides, wind, solar, and geothermal. Each has inherent problems that may be resolved through further research and development. The natural state of these renewable resources, however, have given characteristics. They are generally dilute (weak) in their concentration of energy in comparison to uranium or fossil fuels.

Figure 1-3 Electrical power generation plants: (a) steam; (b) nuclear

Large surface areas are required to produce comparable energy levels, especially tide, wind, and solar sources. Consequently, the necessary equipment is quite large, making the initial cost high.

The continued glut of fossil fuels will incur higher costs due to required additional exploration, drilling, and transportation. Thus as fossil-fuel costs escalate, the renewable sources may eventually become economically competitive. This will be especially true as mass-produced alternative energy systems tend to decrease capital equipment costs. No matter which energy generation source is used, for the purpose of this book it will be assumed that there is unlimited availability.

1-3 POWER CONSUMPTION

Industrialized countries are the world's major energy consumers; an industrial society is synonymous with the consumption of energy. Historically, the consumption-doubling time has been every eight to nine years for electrical energy. This means that during the next nine years more electrical energy will be consumed than in all previous history. This energy must somehow be generated using equally increased amounts of some type of energy source, as described in Section 1-2. The continued application of this philosophy to one's life-style is the real concern of many environmentalist groups and should be the concern of engineers, architects, drafters, designers, and technicians. As noted previously, more and more electrical appliances exist to consume increased amounts of energy. For example, hand-held dryers, which are very common, generally consume 1200 to 1400 W each. Fifteen years ago these "gadgets" were few in number and consumed only 300 to 400 W each. One bright spot to reduce this trend has been the development of microprocessor integrated circuits launching the technical revolution. Television sets, for example, used to consume 400 to 500 W several years ago. A comparable set today uses only 100 to 150 W.

The average person in the United States consumes over 7000 kilowatthours of electrical energy per year. A kilowatthour (kWh) is a unit of energy equal to the use of one 1000-W appliance for 1 hour. Ten 100-W incandescent bulbs operating for 2 hours would equal 2 kWh of use. The annual consumption is somewhat evenly divided among the industrial, commercial, and residential sectors. To help understand the kilowatthour and the cost of electrical energy, values for some common domestic appliances have been given in Appendix Table 1. They have been grouped for use with standard duplex receptacles, or special-purpose circuits. Approximate values have been given for average power ratings and average yearly kilowatthour consumption for several appliances. In addition, the recommended circuit for each special-purpose branch run is included. This is based on 240-V power to all except the 0.75-hp air conditioner, which is 120 V.

Note that the average annual kilowatthour use varies considerably from appliance to appliance. It is a combination of the power drawn by the appliance and the average amount of time it is used. A clock, for example, has low power consumption; however, it is used continuously. Consequently, the annual average kilowatthour consumption is about 8.5 (8.76 kWh per year) times greater than the watts drawn. An electric carving knife, rated at 100 W, shows an annual average kilowatthour consumption at only one-tenth (10) the watts drawn due to intermittent use of the knife.

The cost of electrical energy is based on the kilowatthour. Costs vary significantly in different locales. If an average value is known, 10 cents/kWh, for example, total average annual cost may be determined. To operate a 12-cubic foot refrigerator for 1 year requires approximately 730 kWh (as shown in Appendix Table 1). The cost for electrical energy is: 730 kWh $\times$ 10 cents/kWh = $73. If a frost-free model is used, the annual electrical energy cost increases to $122 (1220 kWh $\times$ 10 cents). To determine the total cost accurately, a rate structure must be obtained for the particular locale. Many of the structures are set so that people are rewarded for higher consumption. This means that the rate per kilowatthour decreases at higher consumption levels. It may change several times during the billing period, so an average value must be used to figure cost.

ASSIGNMENTS

Use an average cost of 10 cents per kWh to solve the following problems. Use average kWh values for the appliances shown in Appendix Table 1.

1-1 Determine the average annual electrical operating cost for a large 4500-W water heater.

1-2 Determine the average annual electrical operating cost for a typical color television set.

1-3 Determine the average annual electrical operating cost for a U.S. citizen.

1-4 Complete a survey listing each electrical-power-consuming device in your residence. Determine the annual kilowatthour of each. Compute the total kilowatthour value, adding only those found in Appendix Table 1. Calculate the combined average annual electrical operating cost.

2

Branch-Circuit Design and Layout

2-1 DIAGRAM FUNDAMENTALS

Electrical power and distribution diagrams are required for residential, commercial, and industrial applications. They differ from other types of diagrams since this is one of the few electrical drawings drawn to scale. The scale usually matches the structural, piping, and heating/ventilating/air-conditioning (HVAC) construction drawings. Normally, it is the plan, or top, view that is shown. It may be accompanied by an elevation view or riser diagram.

Computer-aided design drafting (CAD) is now commonly used for plan-view preparation. A drafter prepares the diagram on CAD using a "layering technique," as shown in Figure 2-1. First a plan view of the building is prepared. This is similar to traditional drafting means, except that it is done on a monitor (a screen resembling a television screen) using an "electronic pencil" (stylus), as shown in Figure 2-2. After the plan has been prepared, it is stored in memory.

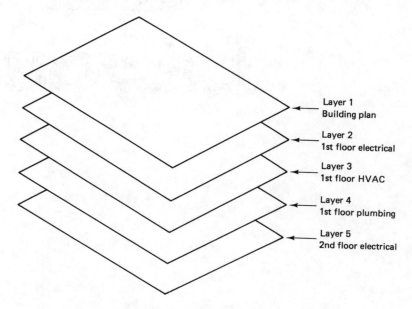

Layer 1
Building plan

Layer 2
1st floor electrical

Layer 3
1st floor HVAC

Layer 4
1st floor plumbing

Layer 5
2nd floor electrical

Figure 2-1 Drawing storage on layers

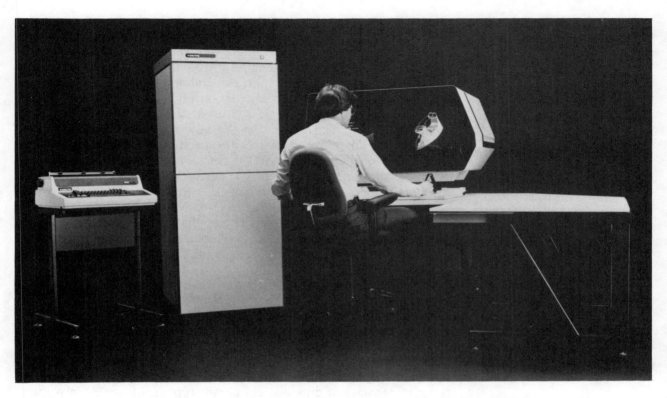

Figure 2-2 Computer-aided drafting station (Courtesy of G.E. Calma)

Each time a different part of the project, such as HVAC, is to be worked on, the plan is recalled and automatically redrawn. That aspect of the job is then worked on. When completed, it too is stored until needed. In this way, the building plan needs to be drawn only once. It can be used repeatedly at the user's discretion. Also, small electrical changes on different floors of a multistory building can be made very rapidly and cleanly. These timesaving methods provide CAD with a significant advantage over traditional drafting. No matter which style is used, however, design drafting principles and codes must be applied. This text covers these standards. The method of drawing preparation is left to the discretion of the reader.

Drawings are represented with the conductors drawn as a single wire. Each single wire is actually two or three conductors plus a grounding wire. As a general rule, all premises wiring shall have a "grounded conductor" (neutral) which normally has gray or white insulation. A "grounding conductor" usually has green or green with a yellow tracer (stripe). Actual system wiring is connected as shown in Figure 2-3. The power and distribution diagram is represented by single-line wiring and symbols. The wires are shown connected to various outlets: lighting fixtures, switches, standard duplex convenience receptacles, and special-purpose receptacles. Each component is represented by a standard symbol. Common standard electrical symbols are shown in Appendix Table 2a. These are used in conjunction with the common architectural symbols shown in Appendix Table 2b.

Branch circuits are governed by *National Electrical Code®* (NEC) Article 210. Most residential and commercial circuits fall under this Section. If a motor, or motor-operated appliance is included in the circuit, Article 430 must also be consulted. If a circuit consists of a motor load only, Article 430 is used. Conductors used for each branch circuit are sized based on the power required by the various applications. The conductor sizes are specified by the American Wire Gage (AWG). The smallest AWG that may be used for residential and commercial applications is No. 14, but usually No. 12 AWG is used. As the numbers decrease, the wire size increases, as shown by the various gages illustrated in Figure 2-4. Note that the larger gages (smaller numbers) are stranded rather than solid. Because a No. 0 gage is approximately 0.32 in. in diameter, it would be difficult to bend during installation if it were solid. The multistranded wire increases flexibility.

A branch circuit is classified according to the maximum-sized fuse or circuit breaker permitted for protection of the circuit. The classifications for standard circuits are 15, 20, 30, 40, and 50 amperes (A). The proper conductor size is selected accordingly. Branch-circuit conductors shall have an ampacity of not less than the rating of the branch circuit and maximum load to be served. The larger the conductor, the greater its ability to carry current safely. The current-carrying capacity of common conductor types is identified in Appen-

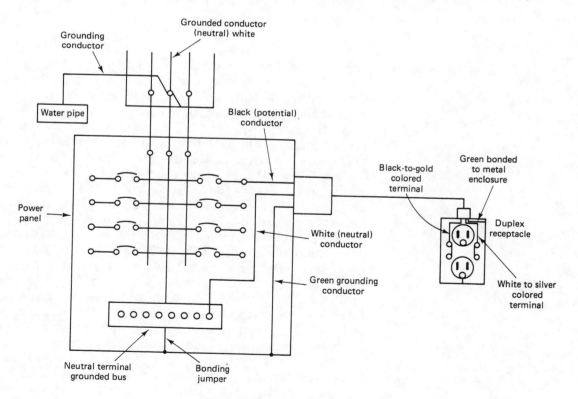

Figure 2-3 Circuit connection

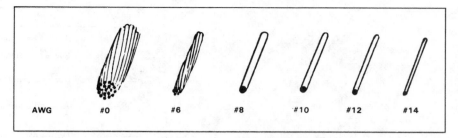

Figure 2-4 Typical American Wire Gages

dix Tables 3a and 3b. From Table 3a, a No. 8 gage copper conductor installed in a raceway or cable can safely carry 50 A with an RHW insulation temperature rise of 75° Celsius (°C). The current-carrying capacity of the same AWG using an aluminum conductor will be only 40 A. Copper is the standard and will be used for all projects in this book. The current-carrying capacity of a conductor used in free air will vary, since wires in raceways are derated because of the heat generated. No. 8 AWG with RHW insulation in free air, for example, can safely carry 70 A (as shown in Table 3b).

Figure 2-5 illustrates a typical plan-view layout for one room of a building. From the symbols shown in Appendix Table 2a, the components used are readily identified as two incandescent lighting fixtures wired to a single-pole switch, several duplex receptacle (convenience) outlets wired together, and a special-purpose outlet for an air-conditioning unit (designated by A/C). The lighting fixtures are connected by conductors that are strung along the ceiling. Receptacle loads are normally connected by wall runs. Often these runs are distinguished by different line styles, such as solid, hidden, or centerlines. The circuits to the room are connected to three separately fused branch runs. These are generally referred to as *branch circuits*, or

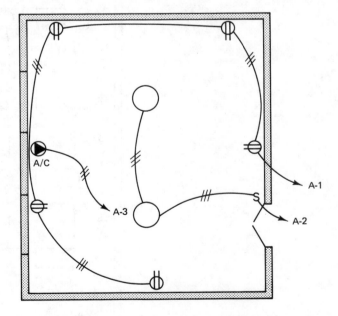

Figure 2-5 Plan view of typical room electrical layout

Symbol	Identification
O	Incandescent lighting fixture
S	Single-pole switch
⊖	Duplex receptacle
▲ A/C	Air-conditioner special circuit
—⫻—	No. 14 AWG
—⫽—	No. 12 AWG

Figure 2-6 Legend for Figure 2-5

home runs, and are designated on the figure by A1, A2, and A3. The slashes that appear intermittently along the wire on each circuit line identify the conductor gage (determined by the ampacity of the circuit) that is used to make each connection. This information is given in a legend which includes the standard symbols used to represent each component or load. A legend for Figure 2-5 would appear as shown in Figure 2-6.

Another common identification technique is to have each slashed line indicate the number of conductors in the cable or raceway. Four short slashes and one long slash, for example, would indicate four conductors with a ground. This system is preferred where more than one branch circuit is pulled through a common cable or raceway.

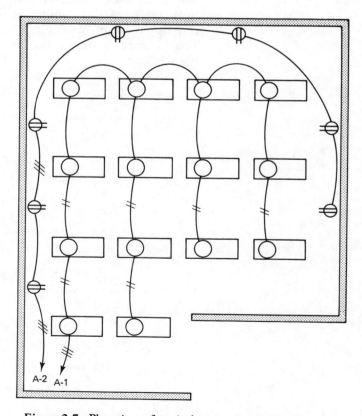

Figure 2-7 Plan view of typical room electrical layout

Since more than one wire gage size would probably exist in the raceway, this version would be required.

It seems redundant to specify standard components in the legend of each layout; however, this is common industrial practice. Premade adhesive aids or rub-ons are often used to save time when preparing standard legends. A legend allows the use of any unusual or special designations. The slashes, for example, can be used to identify any wire gage desired. All wiring methods are presumed to contain a grounding wire as specified by the NEC. Grounding and ground-fault protection have changed significantly in recent years and may be treated as a separate topic by themselves.

Figure 2-7 illustrates a slightly expanded version of Figure 2-5. This is a more typical layout for a room in a commercial building. Note the use of fluorescent fixtures for lighting. This is more common than the use of incandescent lighting because of higher efficiency. Also note that each branch circuit contains either lighting outlets (A1) or receptacle outlets (A2). Except on small circuits, it is best to keep them separate. On larger applications, the lighting and receptacles would be split over two or more circuits each. This way, one breaker trip would not plunge the room into total darkness, nor would all convenience power be lost. The direction of the arrows at the end of each circuit should point toward the general direction of the power panel. Each branch circuit is connected to a fuse or circuit breaker located at the power panel. They are sized according to the power requirements and the American Wire Gage.

2-2 BRANCH-CIRCUIT WIRING

Individual branch circuits, or home runs, separate portions of the wiring circuit from others. Their size is based on the required current draw of individual or grouped loads. As mentioned previously, No. 14 AWG is the minimum conductor size used for residential and commercial purposes and can safely carry 15 A. Some local codes require No. 12 AWG as a minimum, which is also used for convenience outlets. A two-wire feeder supplying two or more branch circuits must be at least a No. 10 AWG conductor. The consumption of electrical energy has been increasing annually. This means that a circuit designed to carry 15 A several years ago may not safely carry the increased current draw of today's higher-wattage appliances. For example, the power consumption of the common hand-held hair dryer has increased over three times during the past several years. For continuous loads, overdesign of a branch circuit is mandatory. Per Article 220 in the NEC, a "continuous load" supplied by a branch circuit shall not exceed 80% of the branch-circuit rating. A continuous load is one in which a full circuit current is expected for 3 hours or more. Examples include lighting for retail stores, offices, traditional drafting rooms, and so on. Except for motor loads, multiple receptacles, and those requiring derating because of different conductors within the same conduit, design the circuits for the commercial projects at 80% of their rated capacity. This means that a circuit using No. 14 AWG will be designed to carry 12 A. For circuits that consume between between 12 and 16 A, No. 12 AWG will be used. The maximum design amperage under normal circumstances for No. 12 AWG is 16 A. This is 80% of the safe limit of 20 A.

Following are the loads for various branch circuits:

circuit rating (A)	receptacle rating (A)	design load (A)
15	15	12
20	20	16
30	30	24
40	40	32
50	50	40

If more than three power and lighting conductors are located in a raceway or cable, or are buried, an additional derating (reduction) is required. This depends on the number of conductors. The following values are the derating percentages of the values in Appendix Table 3:

number of conductors	derating percentage
4–6	80
7–24	70
25–42	60

If these are used, no additional derating for continuous loads is required.

Branch-circuit conductor requirements are listed as follows:

circuit rating (A)	15	20	30	40	50
minimum conductor size (AWG)	14	12	10	8	6

There are a multitude of rules that govern the specifics of electrical circuit installation. The *National Electrical Code*® provides the basic guide that describes national minimum requirements. It is revised and updated every 3 years. Because any of the information given in this section is subject to change, the most recent code book must be consulted before designing branch circuits. This section investigates the basics of branch circuits.

Residential loads are broken into three parts:

1. Lighting (NEC specifies 3 W/ft^2).
2. Convenience outlets (small appliance). At least two circuits (three if laundry area exists) are required.
3. Special purpose (special appliance).

The total number of branch circuits in a residence is based on the number of these three items to be served. Other types of occupancies are broken into the three parts listed above, with the following exceptions:

1. General lighting values are based on the type of occupancy. For example, the NEC specifies a minimum of 2 W/ft^2 allowance for hospitals. Other types of occupancies are listed in Appendix Table 4.
2. General-purpose receptacles are rated at 180 volt-amperes (VA) per duplex convenience outlet. This means that for standard 120-V line voltage, 180 VA ÷ 120 V = 1.50 A, is

required per receptacle. This is not a requirement for residential applications. Good design practice, however, would be to adhere to this rating whenever possible.

3. A "special appliance" is considered to be any load that is not portable.

2-3 SAFETY **Overcurrent Protection** NEC Article 240 governs overcurrent protection of electrical circuits. This is required because the flow of electrical current through a conductor generates heat. An increased flow (amperage) increases the heat generated. If too high, a fire could be started. The safe current-carrying capacity of a conductor is based on an allowable heat generation level. Various American Wire Gages (AWG) having different types of insulation are listed in Appendix

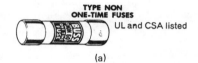

(a)

(b)

Figure 2-8 Fuses: (a) cylindrical; (b) circular

Table 3 by allowable temperature rise. The conductors must be protected by either fuses or circuit breakers. A cylindrical fuse is shown in Figure 2-8(a). Circular types are shown in Figure 2-8(b) connected in a typical application. Fuses are set no higher than the ampacity rating of the conductors. For example, a No. 00 AWG in conduit having type T (thermoplastic) insulation has an allowable current capacity of 145 A at a 60°C (140°F) temperature rise. A fuse or circuit breaker, if standard, would be used. If not, the next higher standard is used. Standard fuse sizes are 1, 3, 6, 10, 15, 20, 25, 30, 35, 40, 45, 50, 60, 70, 80, 90, 100, 110, 125, 150, 175, 200, and 225 A. Since 145 A is not standard, 150 A would be used. Typical overcurrent protection for branch-circuit conductors are as follows:

current rating (A)	15	20	30	40	50
minimum conductor size (AWG)	14	12	10	8	6
overcurrent protection (A)	15	20	30	40	50

Circuit Breakers   Circuit breakers, like fuses, have standard trip settings. Typical circuit breakers are shown in Figure 2-9. The settings are the same except for 1, 3, 6, and 10 A. The minimum residential circuit breakers is 15 A. Upon system failure, the circuit breaker trips, cutting power to the system and thus preventing the conductors from overheating. After correction of the malfunction, it can be manually reset for use again, and power is restored to the system. A fuse cannot be reset; the fuse element or the fuse itself must be replaced. The circuit breaker will have a rating no less than the trip rating. This is referred to as the *frame size*. Standard frame sizes in-

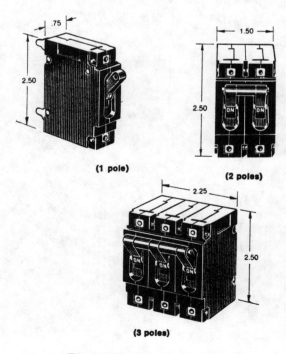

(1 pole)

(2 poles)

(3 poles)

Figure 2-9 Circuit breakers

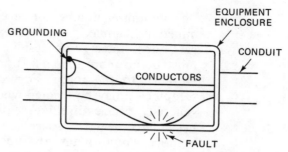

Figure 2-10 Typical fault

clude 50, 100, 225, 400, and 600 A. Thus, in the earlier example of 150 A to trip, a 225-A frame would be used.

For three-phase protection, an overcurrent device is placed in each of the three current-carrying conductors (phase legs). No overcurrent device is placed in the grounded neutral line.

Grounding All premises wiring must be grounded in accordance with NEC Article 250. There are two types of grounding:

1. Circuit and system grounding
2. Equipment grounding

Refer to Article 250, which also provides reference to a multitude of other Articles and Sections of the book for specific application. In general, however, proper grounding protects a person from accidental shock in case of a fault, as shown in Figure 2-10. Typical grounding methods are shown in Figure 2-3.

Besides the requirement that all receptacle circuits be grounded, ground-fault interruption (GFI) must be provided in certain areas. A ground-fault interrupter is a device that deenergizes a circuit when current to ground exceeds a predetermined value. For all 120-V, single-phase residential applications, ground-fault interruption protection must be provided in:

1. Bathrooms
2. Garages
3. Outdoors with grade-level access

A branch circuit must be provided for this purpose.

2-4 STANDARD RECEPTACLES

For residential application, receptacle outlets shall be installed in the following: kitchen, family room, dining room, living room, parlor, library, den, sun room, bedroom, recreation room, or any similar rooms.

Standard duplex wall receptacles for residential and commercial application are rated at not less than 180 VA each. Common design practice uses 1.5 A each for sizing a branch circuit for 120-V systems.

The minimum number of receptacle outlets in a building is determined by the maximum distance between plugs. The *National Electrical Code®* states: "For [residential dwellings] application, receptacles must be no more than 12 ft apart and no more than 6 ft from any entrance to the room. For commercial and industrial application, receptacles may be located where needed." The projects in this book generally utilize dwelling requirements. Thus if 20 receptacles are needed to meet the space requirements in a building, good design practice would dictate their combined rating to be 30 A. Since No. 12 AWG service is normally used for standard "small portable appliance" service, two 20-A circuits would be required. A branch circuit contains two or more outlets, and a special-purpose circuit contains only one.

Specific receptacle requirements include:

1. Continuous wall locations 12 ft apart (maximum)
2. All parts of wall space greater than 6 ft (measured horizontally)
3. A receptacle for two openings or doorways greater than 2 ft apart
4. A receptacle for counter space greater than 1 ft long
5. A receptacle within 6 ft of all parts of a room divider

These requirements are illustrated in Figure 2-11. The 10 receptacles shown would be connected to one 20-A branch circuit. Other common-used residential provisions include:

1. Laundry room receptacles are to have at least one 20-A branch circuit. Receptacles are to be placed within 6 ft of the appliance(s). No other receptacles in other rooms may be attached to it.

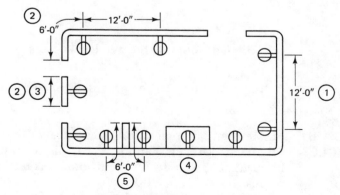

Figure 2-11 Receptacle layout

2. At least one convenience outlet is required in bathrooms, outdoors, and basements. The bathroom and outdoor receptacles are connected to a ground-fault interruption circuit.

3. Two or more receptacle circuits are required in a kitchen area. These circuits are also permitted to supply other receptacles in other rooms.

Commercial and industrial outlets must be based on 180 VA per duplex outlet. Above 10 kilowatts (kW), however, the demand factor reduces to 50%. Special provisions are made for other commercial factors. For kitchen equipment, the demand factors are as follows:

number of units of equipment	demand factor (%)
1	100
2	100
3	90
4	80
5	70
6 and over	65

Other provisions for receptacles will be used, as needed, in the design projects.

2-5 INTERIOR LIGHTING

Interior lighting design and layout in a residential application is generally not significant since it can easily be modified. The illumination level can be increased by using a larger-wattage bulb (up to the ampacity of the conductors), or by changing the location of movable lamp fixtures. Remember to allow 3 W/ft^2 (refer to Appendix Table 4) when sizing branch circuits. Also, do not include the floor area of open porches, garages, and spaces not to be used. In a commercial or industrial application it is critical that the design and location be as accurate as possible the first time, since they are generally permanent. For example, in exterior lighting, it would be expensive to change the locations of lamp poles once installed. This section will cover illumination levels, design calculations, and the layout of interior lighting.

Minimum lighting loads for various occupancies are specified in Appendix Table 4. A computed load of these minimum values must be used. Illumination levels, however, depend on several factors: the amount of light (strength) of the source, the amount of light reaching the surface, and the losses between the source and the surface. The amount of light at the source is measured in units of lumens. The lumens per watt (lm/W) differ depending on the type of lighting used. The three common interior sources used are the following:

type	approximate lm/W
Fluorescent	80
Mercury vapor	50
Incandescent	20

Even though these values vary with wattage, one can quickly see the differences in efficiency. A fluorescent bulb produces approximately four times the illumination of that of an incandescent bulb of the same wattage. Various outputs of common lamps are listed in Appendix Table 5 and are illustrated in Figure 2-12.

The amount of light at the surface is measured in footcandles (fc). Different applications require different lighting levels. For example, a room where continuous reading or close detail work is done may require a level in the 100-fc range. On the other hand, a storeroom may require only 10 or 20 fc. Recommendations of lighting levels have been made by different groups, but the most widely accepted are those of the Illumination Engineering Society (IES). The minimum levels, which conform closely to IES recommendations, will be used for the solutions to the projects in this book and are given in Appendix Table 6.

All of the light emitted by the source is not available at the surface. One cause is aging; as lamps are used, the lumen output decreases. For example, the illumination of a 40-W fluorescent fixture initially yielding 3100 lm will drop to less than 2950 lm after 40% of its rated life has passed. Rated life varies with the type of lamp and the manufacturer. Also, as the lamps, walls, ceilings, and floors wear and gather dirt, the amount of light reaching the surface decreases. The term used to take this into account is the *maintenance factor* (M.F.). The following values will be used for lighting design:

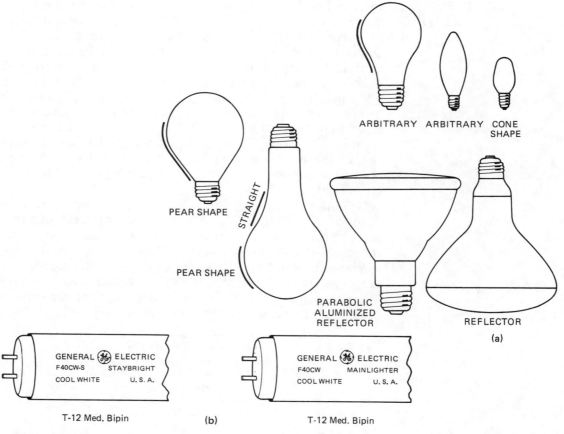

Figure 2-12 Typical lamps: (a) incandescent; (b) fluorescent (Courtesy of General Electric Corp.)

- Good maintenance   0.70–0.80
- Medium (average)   0.65
- Poor maintenance   0.50–0.60

The exact values will vary according to specific application; however, a reasonable estimate can be made. For example, a dirty environment may have an M.F. of 0.50, thus reducing the available light by half.

The other major factor affecting lighting efficiency is the *coefficient of utilization* (C.U.). This takes into account the room shape, including the distance from the lamps to the surface, type of luminaire, and the dimensions (length times width) of the room. The illumination pattern of a lamp in different fixtures will vary, as shown in Figure 2-13. Also affecting C.U. are the color of the walls, floor, and ceiling. A lighter wall surface will generally reflect more light, thus improving the efficiency. An average value often used for the C.U. is 0.60. This number means that 40% of the illumination has been lost due to this factor.

If the type of luminaire (lighting fixture) and room size information are known, the C.U. can be determined more closely. Appendix Table 7 can be used to determine the C.U. using the common types of luminaire shown in Figure 2-13. For example, a room 22 ft wide by 55 ft long has the lighting fixtures located 6.0 ft from the work surface with good wall and ceiling reflectance. Using part (A) of Appendix Table 7, the intersection of the room size and distance shows a room cavity ratio (CR) number of 2.0. This information is coupled with the 80% ceiling reflectance and 50% wall reflectance in part (B) of Appendix Table 7 to yield a coefficient of utilization of 0.60. Appendix Table 7 will be used to determine the C.U. for all projects in this book unless otherwise instructed.

If all data for a design project are known, the number of lighting fixtures required can be determined using the following lumen formula:

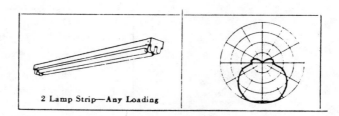

2 Lamp Strip—Any Loading

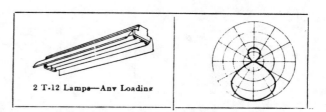

2 T-12 Lamps—Any Loading

Figure 2-13 Common luminaire lighting pattern

$$fc = \frac{(\text{lm/lamp})(\text{no. lamps})(\text{C.U.})(\text{M.F.})}{\text{surface area (ft}^2)}$$

where fc = desired footcandle lighting level
  (Appendix Table 6)
lm/lamp = output rating of the source
  (Appendix Table 5)
no. lamps = minimum required number of bulbs
C.U. = coefficient of utilization
  (use 0.60 for average value, or Appendix Table 7)
M.F. = maintenance factor
  (use 0.65 for average value)
surface area = size of the surface (square feet)

For example, if a 22 ft by 55 ft room is used for general office purposes and utilizes 40-W (3100-lm fluorescent bulbs), the required number of lamps is

$$80 = \frac{(3100)(\text{no. lamps})(0.60)(0.65)}{22 \times 55}$$

no. lamps = 80.06 (use 80 lamps)

Many common fixtures contain two or four lamps. If two-lamp fixtures are used, then 40 are required. The result from the lumen formula may be rounded to an even integer so that all 40 fixtures are complete with bulbs. Also, the result may be varied to accommodate the best layout. Forty fixtures would probably yield a better lighting pattern than 39 fixtures. For example, five rows of eight or four rows of 10 might be used. In a grid-type channel ceiling, the pattern would have to complement the ceiling.

During fixture layout, the goal is to obtain a uniform pattern of illumination. Figure 2-14 illustrates a portion of a typical general office layout. The fixtures are placed closer to the walls than to each other, since a point anywhere inside receives illumination from more

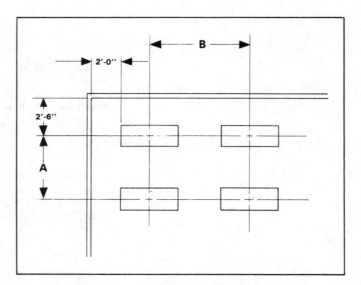

Figure 2-14 Partial fluorescent lighting layout

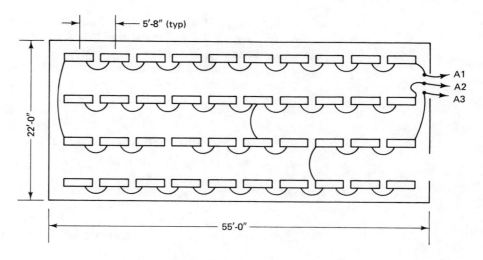

Figure 2-15 Fluorescent lighting layout

than one fixture. The walls do not receive this amount of shared lighting and if they are of a dark color or rough texture, more absorption occurs. For layouts with a minimum level of 50 fc, 2 ft 6 in. will be used between the wall and the center of the fixture's narrow side, and 2 ft 0 in. between the wall and the edge of the fixture's long side. Lower illumination levels will necessitate a greater distance.

For the office building example above, four rows of 10 fixtures will be used. This will make dimensions A and B of Figure 2-14 between centers of the horizontal and vertical rows of fixtures close to 5 ft 8 in. each. Whenever possible, these dimensions should be close or equal. The fixtures are drawn to scale. For the 40-W bulbs used, the fixture size is approximately 8 in. wide by 50 in. long. On large-scale reductions, however, they may be drawn larger for clarity. A plan view of the office overhead fluorescent lighting layout is shown in Figure 2-15. Note that three branch circuits are specified. These are identified as A1, A2, and A3. The identification refers to fuse (or circuit breaker) circuits 1, 2, and 3 located in power panel A. Interior lighting branch-circuit runs are normally made with No. 14 or No. 12 AWG (No. 12 in this case) conductors. These smaller gages are easier to work with during assembly, when bending and so on. Also, they preclude the necessity of using heavy-duty lampholders.

2-6 EXTERIOR LIGHTING

Interior lighting (Section 2-5) should be studied before this section. Many of the principles are the same and will not be expanded on. As with interior lamps, different types of lighting are available. These are referred to as high-intensity-discharge (HID) lamps. Three types of light sources comprise the HID lamp category:

type	approximate lm/W
Mercury vapor	40–60
Multivapor (metal halide)	80–100
High-pressure sodium vapor	100–130

Various HID lamps are shown in Figure 2-16.

Again, as in the various types of interior lighting, significant differences in efficiency can easily be seen. Each of the bulb types has features that will be used in different applications. Typical applications include:

- Area and floodlighting (commercial and industrial buildings)
- Sporting stadiums, arenas, tennis courts, and so on
- Parking and auto lots
- Roadway lighting

Except for mercury vapor, the applications of HID lamps are exterior situations. In particular, high-pressure sodium vapor, termed

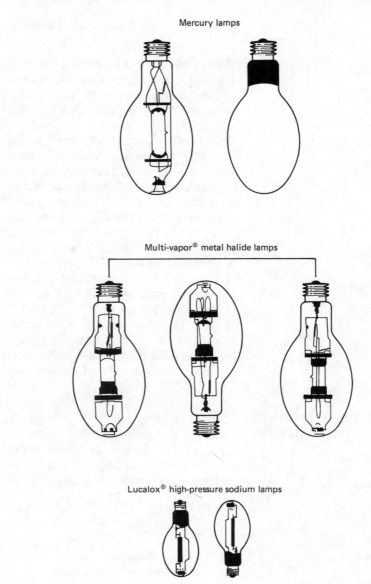

Figure 2-16 High-intensity-discharge lamps (Courtesy of General Electric Corp.)

Lucalox® by the General Electric Corp., is highly efficient. It is, however, not pleasing to the eye and thus is used essentially for parking lots or roadways. It is, in fact, the dominant type used during new construction. To provide increased efficiency, existing parking lot and roadway mercury vapor lighting is being converted to high-pressure sodium vapor lighting.

Lamp sizes range between 50 and 3000 W. These include:

type	wattage range	normal sizes (W)
Mercury vapor	50–3000	100, 175, 250, 400, 1000
Multivapor	400–1500	400, 1000
Sodium vapor	150–1000	175, 250, 400, 1000

For parking lot and roadway design in this chapter, 400- to 1000-W lamps will be used. A typical luminaire for this purpose is shown in Figure 2-17. For walkways the lamp size will be 175 or 250 W, similar to the luminaire shown in Figure 2-18. The mounting height or distance from the source to the surface ranges between 15 and 20 ft for 175- to 250-W lamps, to 25 to 30 ft for 400- to 1000-W lamps. Mounting heights are high since most applications will require broad coverage at low footcandle levels. Various outputs of common HID lamps are shown in Appendix Table 8. As previously indicated, lumen output of all lamps will decrease with use. A graph showing this decay for HID lamps is illustrated in Figure 2-19.

The lumen formula presented in the interior lighting (Section 2-5) is used by some manufacturers. Some difficulty, however, is encountered in determining the maintenance factor and coefficient of utilization. General maintenance and room shape/color features are not factors. The lamps are cleansed by nature and the world is the room. Most manufacturers use methods that combine these factors. Appendix Table 8 includes these factors in the values presented.

Figure 2-17 Parking lot and roadway luminaire

Figure 2-18 Walkway luminaire

The footcandle levels for outdoor lighting are normally quite low. Installations on roadways, parking lots, and walkways, and for safety and security purposes, require illumination levels only in the range 0.5 to 5.0 fc; the levels will normally be between 1.0 and 2.0 fc. For example, using Appendix Table 8, a walkway application using one 175-W mercury vapor lamp will yield 1.0 fc at a surface distance of 5 ft from the lamp in any direction. Moving 15 ft away from the lamp in any direction will reduce the level to 0.4 fc. Because of shared lighting, two lamps installed 30 ft apart will yield a level of approximately 0.8 fc midway between the lamps. Other common footcandle levels, with the approximate spacing or surface area per lamp are given in Appendix Table 8. The values in the table include an allowance for mounting height.

The lamps are mounted with supports that vary depending on the application. Roadway lighting is normally single-arm, pole-mounted. The vertical pole is either sunk into the ground or is set on a pedestal with anchor bolts, as shown in Figure 2-20. The conductors run inside the pole up to the fixture. This part of the installation is

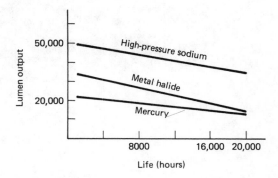

Figure 2-19 Lumen decay

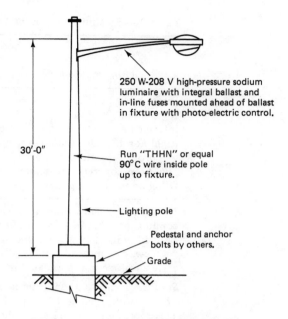

250 W-208 V high-pressure sodium
luminaire with integral ballast and
in-line fuses mounted ahead of ballast
in fixture with photo-electric control.

Run "THHN" or equal
90°C wire inside pole
up to fixture.

30'-0"

Lighting pole

Pedestal and anchor
bolts by others.

Grade

Figure 2-20   Roadway lighting application

done by others and is not shown on the drawing. For the projects in
this chapter, underground wiring will be sized and provided at each
pole. Cable suitable for direct burial, such as type UF, may be used.
Refer to NEC Sections 339 and 225 for specifics. Parking lot or area
lighting can be either single- or double-arm pole-mounted. Platform,
stair, or working lights, also done by others, will be rail-mounted.
This installation will not be shown on the drawing.

Layout of the illumination pattern is similar to the one presented
for interior lighting, with two exceptions. Walls need not be considered, and the fixtures are spaced much farther apart. A parking lot
layout using mercury vapor lamps and providing a footcandle level of
1.5 is shown in Figure 2-21. The direct buried cable or conduit is installed to meet minimum cover requirements. For installations under
600 V, the minimum required depth is specified by NEC Article
330-5 and shown in Figure 2-22.

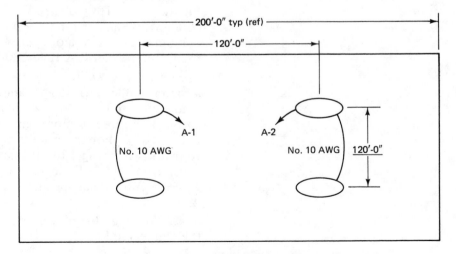

Figure 2-21   Parking lot layout

Table 300-5   Minimum Cover* Requirements, 0 to 600 V, Nominal

wiring method	minimum burial (in.)†
Direct buried cables	24
Rigid metal conduit	6
Intermediate metal conduit	6
Rigid nonmetallic conduit approved for direct burial without concrete encasement	18
Other approved raceways‡	18

*Cover is defined as the distance between the top surface of direct buried cable, conduit, or other raceways and the finished grade.

† For SI units: 1 in. = 25.4 mm.

‡ Raceways approved for burial only when concrete encased shall require a concrete envelope not less than 2 in. (50.8 mm) thick.

Source: Reprinted with permission from NFPA70-1984, *National Electrical Code®*, Copyright © 1983, National Fire Protection Association, Boston, Mass. This reprinted material is not the complete and official position of the NFPA on the referenced subject, which is represented only by the standard in its entirety.

Figure 2-22 Minimum cover requirements

2-7 SPECIAL-PURPOSE LOADS

Large power-consumption non-portable devices require special-purpose loads (individual branch circuits). These are used for a single purpose. Generally, a No. 14 AWG (15 A of overprotection) is the minimum size conductor. Exceptions to this are ranges with a 8.75-kW rating (or greater), requiring a minimum overprotection of 40 A. Normal special-purpose load practices will be given below, together with the appropriate NEC Article.

The common loads are categorized as follows:

1. *Ranges.* Demand loads for residential electric ranges and similar appliances will vary depending on number and power consumption. The feeder is based on the table shown in Figure 2-23. Thus a 12-kW range is figured at 8 kW per column A. Smaller ranges use a percentage of the rating per column B or C. For a one 5-kW load, 80% or 4 kW would be used. Refer to NEC Article 220-19 for exceptions.

2. *Commercial kitchen equipment.* For commercial kitchens, a variety of electrical appliances will be utilized. These include such items as dishwasher heaters, water heaters, and so on. For three or more units, the demand factor will drop below 100%. For three units the demand factor is 90%, for four units it changes to 80%, and for five units it changes to 70%. Each unit thereafter drops the feeder demand factor to 65%. Additional specifications are found in NEC Article 422.

3. *Space heating.* Fixed electric space heating uses 100% of the connected load.

Table 220-19 Demand Loads for Household Electric Ranges, Wall-Mounted Ovens, Counter-Mounted Cooking Units, and Other Household Cooking Appliances over 1¾ kW Rating. Column A to be used in all cases except as otherwise permitted in Note 3 below.

number of appliances	maximum demand (kW) (see notes): column A (not over 12 kW rating)	demand factors (percent) (see Note 3) column B (less than 3½ kW rating)	column C (3½ to 8¾ kW rating)
1	8	80	80
2	11	75	65
3	14	70	55
4	17	66	50
5	20	62	45
6	21	59	43
7	22	56	40
8	23	53	36
9	24	51	35
10	25	49	34
11	26	47	32
12	27	45	32
13	28	43	32
14	29	41	32
15	30	40	32
16	31	39	28
17	32	38	28
18	33	37	28
19	34	36	28
20	35	35	28
21	36	34	26
22	37	33	26
23	38	32	26
24	39	31	26
25	40	30	26
26-30	15 kW plus 1 kW for each range	30	24
31-40		30	22
41-50	25 kW plus ¾ kW for each range	30	20
51-60		30	18
61 and over		30	16

Note 1: Over 12 kW through 27 kW ranges all of same rating. For ranges individually rated more than 12 kW but not more than 27 kW, the maximum demand in column A shall be increased 5% for each.

Source: Reprinted with permission from NFPA70-1984, *National Electrical Code*®, Copyright © 1983 National Fire Protection Association, Boston, Mass. This reprinted material is not the complete and official position of the NFPA on the referenced subject, which is represented only by the standard in its entirety.

Figure 2-23 Electric range demand loads

4. *Clothes dryer.* The load for a residential electric clothes dryer is 5000 W or the nameplate rating, whichever is larger. When using more than four dryers, refer to NEC Article 220-18 for a reduction in the demand factor.

5. *Motors.* Use NEC Article 430. In general, however, the branch-circuit conductor supplying a single motor has an ampacity of at least 125% of full-load current. This is true for motors operating at a level of 600 V or less. The over-protection device will trip at 125% (maximum) of the full-load current. For multiple-motor circuits, the feeder is sized at 125% of the largest motor load plus the sum of the other motor loads:

$$\text{feeder ampacity} = I_{m1} \times 1.25 + I_{m2} + I_{m3} + \cdots$$

The full load currents of single- and three-phase motors are given in Appendix Tables 9a and 9b. A 2-hp single-phase motor circuit operating at 230 V draws a full-load current of 12 A. The branch-circuit conductors are sized for a 12 A × 1.25 = 15 A capacity. An example of feeder ampacity requirements for several motors is as follows. Three three-phase motors operate at 230 V. Two are 5-hp motors and the other is 3 hp. Using Appendix Table 9b, the full-load currents are 15.2, 15.2, and 9.6 A. The feeder ampacity will be

$$15.2 \times 1.25 + 15.2 + 9.6 = 43.8 \text{ A}$$

6. *Air conditioning/refrigeration.* NEC Article 440 specifies the provisions for air-conditioning and refrigeration equipment. Since these devices are normally motor-driven, compliance is to the motor section (Article 430).

2-8 BRANCH-CIRCUIT DESIGN

This section covers sizing of individual branch circuits resulting from electrical loads. The electrical consumption will be the type developed by loads described in Sections 2-4, 2-5, and 2-7. Compliance is to NEC Articles 210 and 220. Other Sections that affect specific-purpose branch circuits are listed in Article 210.

Receptacle Load   Two rooms of a small building are represented by the plan shown in Figure 2-24. The use of at least 20 duplex convenience receptacles based on residential application is required. As mentioned previously, each receptacle is based on a minimum of 180 volt-amperes (VA) derived from:

$$\text{VA} = I \times E$$
$$180 \text{ VA} = 1.5 \text{ A} \times 120 \text{ V}$$

Normally, No. 12 AWG is used for receptacle branch circuits rated at 20 A, as specified in Appendix Table 3. The maximum design load for this is 16 A (80%). Thus

$$I_{\text{total}} = I/\text{outlet} \times \text{no. outlets}$$

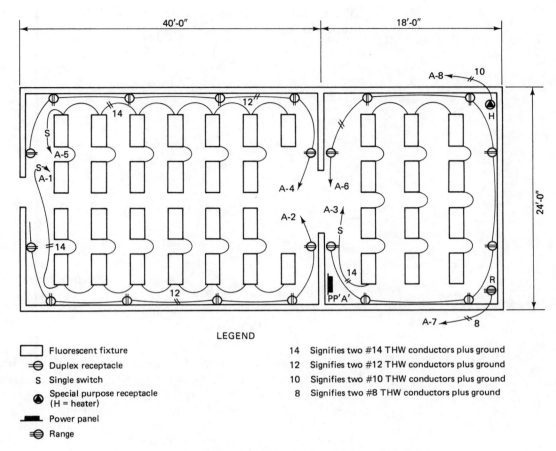

LEGEND

▭	Fluorescent fixture	14	Signifies two #14 THW conductors plus ground
⊖	Duplex receptacle	12	Signifies two #12 THW conductors plus ground
S	Single switch	10	Signifies two #10 THW conductors plus ground
⬤	Special purpose receptacle (H = heater)	8	Signifies two #8 THW conductors plus ground
▬	Power panel		
⊜	Range		

Figure 2-24 Branch circuit plan view

or

$$\frac{I_{total}}{I/\text{outlet}} = \text{no. outlets (maximum)}$$

$$\frac{16 \text{ A}}{1.5 \text{ A/outlet}} = 10.7$$

or a maximum of 10 receptacles may be used for each 20-A branch circuit. A minimum of two branch circuits is required. For commercial applications it is best to use fewer than the maximum number of receptacles per 20 A circuit, for example, eight. Thus three branch circuits are used. In Figure 2-24, three 20-A circuits are shown for receptacle loads. These are designated as A4, A5, and A6 and refer to power panel A, circuits 4, 5, and 6.

Lighting A general lighting load may be computed quickly by the use of Appendix Table 4. If the plan shown in Figure 2-24 is either a residence or a commercial store, use 3 W/ft². The minimum load is based on the total floor area using outside dimensions:

$$\text{floor area} = 58 \times 24 = 1392 \text{ ft}^2$$

$$\text{load} = 1392 \text{ ft}^2 \times 3 \text{ W/ft}^2 = 4176 \text{ W}$$

The minimum required number of branch circuits may then be determined as follows:

1. Based on 15-A branch circuits at 120 V and a continuous load, use 12 A (80%) for design purposes:

$$P = E \times I$$

where P = power (watts)
E = electromotive force (volts)
I = current (amperes)

watts/circuit = 120 V $\times$ 12 A = 1440 W/circuit

2. The minimum number of circuits becomes

$$\frac{\text{load}}{\text{W/circuit}} = \text{no. circuits (minimum)}$$

$$\frac{4176 \text{ W}}{1440 \text{ W/circuit}} = 2.9 \text{ or } 3 \text{ 15-A circuits (mimimum)}$$

An alternative method to determine lighting-load branch circuits may have to be used. If a desired footcandle level has been established, the size and number of bulbs must first be computed. The method described in Section 2-5 is used for this purpose. Next, the current is determined. For incandescent lighting and pure resistance loads, the current drawn can be determined by

$$P = E \times I$$

If the voltage level, the total number of lamps, and each wattage are known, the current can be calculated. For example, if fourteen 100-W incandescent lamps are to be part of one branch circuit:

$$P = E \times I$$

$$(100)(14) = 120 \text{ V} \times I$$

$$I = \frac{1400}{120} = 11.7 \text{ A}$$

A branch circuit with No. 14 AWG and a 15-A fuse or breaker would be selected since the maximum design load is 12 A.

Normally, lighting circuits operate with No. 12 or 14 AWG at nominal voltages of 120, 208, 240, or 277. A 120-V circuit will be used to determine interior lighting circuits for all projects in this book.

The method described above cannot be used for sizing inductive lighting circuits. Fluorescent lighting is inductive since it contains ballasts, transformers, and autotransformers. The ballast limits the current and provides starting. Consequently, a power factor must be taken into account for this additional load. A general rule is to add 25% of rated bulb wattage to compensate for this. Thus a 40-W bulb actually consumes 50 W, an 80-W bulb consumes 100 W, and so on.

This also means that fluorescent lighting is not actually four times more efficient than incandescent, as indicated earlier; it is closer to three times more efficient.

The branch-circuit determination for seventy-six 40-W lamps may have alternate solutions. First, the current can be determined by

$$P = E \times I$$
$$(76)(40 \times 125\%) = 120 \text{ V} \times I$$
$$I = \frac{3800}{120} = 31.9 \text{ A} = 32 \text{ A}$$

Based on a continuous load (80%), the minimum number of circuits becomes either

two 20-A No. 12 AWG circuits
[16 A + 16 A = 32 A (maximum design level)]

or

three 15-A No. 14 AWG circuits
[12 A + 12 A + 12 A = 36 A (maximum design level)]

If the three 15-A branch circuits and two bulbs per fixture are chosen, a lighting layout similar to the one shown in Figure 2-24 might be used. The three branch circuits are designated by A1, A2, and A3. These specify that their location is power panel A, circuits, 1, 2, and 3. Note that each is controlled by a switch provided at an entrance to the area.

Several options to connect the loads are possible. The circuits may be split in a more random pattern. The lighting circuits, for example, may be connected to every other row. This would provide better intermittent lighting at those times when only one switch circuit is operating. Also, connecting receptacles in the same room on different branch circuits ensures that all power will not be lost to any one room by one breaker trip.

Special-Purpose Circuits Special branch circuits and receptacles are necessary for any load that exceeds the wattage of the "small appliance" group, which normally includes ranges, heaters, air conditioners, motors, and others described in Section 2-7.

Ranges under 9 kW are rated by an 80% demand factor, as shown in Figure 2-23. For example, 8.75 kW $\times$ 0.8 = 7.0 kW demand load. If the range operates at 240 V, the current drawn is

$$I = \frac{P}{E}$$
$$= \frac{7000}{240} = 29 \text{ A}$$

One No. 8 AWG 40-A special-purpose receptacle (next higher standard size) on one branch circuit will be provided. *Note:* 40 A is the minimum size allowed for ranges by NEC.

If an electric water heater draws 5000 W and operates at 240 V,

$$I = \frac{P}{E}$$

$$= \frac{5000}{240} = 20.8 \text{ A} = 21 \text{ A}$$

Since the branch-circuit rating must not be less than 125% of current drawn,

$$I_{design} = I_{drawn} \times 1.25 = 26.25 \text{ A}$$

Use a special-purpose receptacle and a No. 10 AWG branch circuit with a 30-A breaker (next higher standard size).

The current drawn by various motors differs. It depends on the type of motor, the manufacturer, and the speed and efficiency of the motor. Appendix Tables 9a and 9b show typical current ratings at various operating voltages. Use this table to size branch circuits for motors required on the projects. The table may also be used to size an accompanying transformer. Table 10 provides additional information regarding motor circuits. After branch sizing is understood by the reader, this table may be used as a quick, handy, reference guide.

The branch-circuit sizing presented thus far is for standard conductor run length. The power and distribution layout for each of the branch-circuit examples determined in this section is illustrated in Figure 2-24. The range and electric water heater special-purpose circuits are indicated as A-7 and A-8.

For longer runs over approximately 75 ft, negative effects will occur. The conductor has resistance; thus a voltage drop is developed, based on Ohm's law, which is

$$E = IR$$

where E = electromotive force (volts)
 I = current (amperes)
 R = resistance of the conductor (ohms)

If the resistance of the conductor is known, the drop in voltage can be determined. Normally, this drop is held to within 3%. An excessive drop will cause electrical devices to function improperly and shorten their life spans significantly. To prevent this, a larger wire gage is used, since a larger cross section has less resistance. Consult Appendix Table 11 when designing long-run branch circuits. For example, 20 A is the maximum safe current for a branch circuit. With no derating, a No. 12 AWG conductor can be used. If the length of run is 200 ft, however, this changes the conductor size. Refer to the 20-A column in Appendix Table 11. Find the closest value above 200 ft in the column. Read across this row to the size. The recommended conductor size of this case is No. 6 AWG. The resistance of common gage conductors found in Appendix Table 12 can be used to determine the actual voltage drop for the applications in this book. Other resistance values for any gage conductor may be found in any electrical theory text.

An additional consideration is necessary for long runs having conductor splices. Breaking one conductor into two requires the use of a junction box. Even if a break into two branch lines is not made, junction boxes are used periodically along the run. This provides a place to pull the wire through, referred to as a *pull box*. Pull boxes are provided so that a wire would not have to be pulled through conduit, walls, or underground at excessive lengths. A run of 300 ft, for example, would be unacceptable and the pulling force required would probably exceed the conductor tensile strength. When using pull boxes, certain requirements must be met. For example, the length of the box must be eight times longer than the diameter of the conduit. This is a general requirement that must be adhered to. It provides the electrician with adequate room to pull the wire and make the connection. This rule is also true for connections requiring changes in direction (example: 90° bend). Thus many special and additional requirements based on NEC must be adhered to by the electrician after the design, conduit selection, and layout have been made.

2-9 COMMERCIAL DESIGN

The electrical plan for a service store commercial application is shown in Figure 2-25. The electrical symbol legend identifies each of the loads used and corresponds to Appendix Table 2a. Note that two power panels have been specified. Panel A indicates all receptacle and lighting branch circuits. Panel P is indicated on all special-purpose circuits. This includes air conditioning, heaters, motors, and other large loads. All layouts will contain a title block to be located in the lower right-hand corner of the drawing. A typical title block to be used for the projects in this text is given in Appendix 13.

ASSIGNMENTS

Project work in Chapter 2 should be accompanied with the NEC.

2-1 Single-Room Load Requirement.

Given: A residential family room with one entrance has 18 ft × 25 ft floor dimensions.

Required:

    a. Determine the number of duplex wall receptacles that would be used.

    b. Compute the lighting and receptacle branch-circuit requirement. Use Appendix Table 4 for lighting.

2-2 Interior Lighting for an Office.

Given: Calculate the interior lighting requirement of the computer-assisted design and drafting (CAD) office shown in Figure P2-2. The illumination level is to be 60 fc, and the operating voltage level is 120 V. Use 0.65 each for the coefficient of utilization and maintenance factor. The lighting will be fluorescent, two bulbs per fixture, using 40-W 3100-lm bulbs.

Required: Include the following information on the layout:

    a. Overhead lighting pattern. Show each branch circuit run using No. 14 or 12 AWG only. Split the circuits so that the entire room does not have to be illuminated.

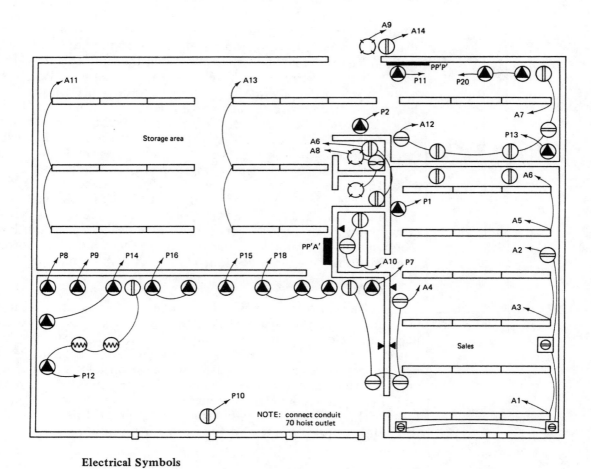

Electrical Symbols

⊘	Incandescent-light fixture on ceiling
▭	Fluorescent-light fixture on ceiling
⊕	Duplex receptacle on wall
⊟	Duplex floor receptacle
◀	Telephone outlet
▬	Panel board, wall mounted
$	Single-pole toggle wall switch

Single circuit to panel board   A = lighting circuit
                                        P = power circuit
Example A3 = No. 3 lighting circuit

▲ Special-purpose power outlet

〰 Floor receptacle — Class I explosion proof
for wheel alignment machine

No wiring less than No. 12 AWG

Figure 2-25   Commercial plan

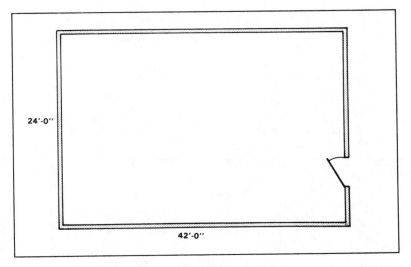

24'-0"

42'-0"

Figure P2-2

b. Legend, including the component symbols, conductor sizing, number of fixtures required, and the number of fixtures used.

2-3   Residential Floor Plan.

Given: A layout of the electrical plan for a small residence is shown in Figure P2-3.

Required: Answer the following questions:

a. How many duplex wall receptacles are used?

b. How many receptacle branch circuits would be used?

c. What are the special-purpose circuit(s) used for?

2-4   Lighting and Circuit Layout for a Residence.

Given: A plan layout for a residence is shown in Figure P2-4. Use the NEC requirements outlined in this chapter for the lighting and receptacle branch circuits. In addition, include provisions for a 12-kW range in the kitchen area.

Required:

a. Determine the number of branch circuits required.

b. Prepare a layout of the electrical plan.

c. Provide a legend, including the component symbols and conductor sizing.

2-5   Exterior Illumination.

Given: A parking lot for a commercial store is 350 ft wide × 600 ft long.

Required:

a. Determine the required number of 400-W luminaires for 1.0-fc illumination.

b. Using No. 10 AWG conductors, determine the number of branch circuits required.

c. Prepare a layout of the electrical plan. Include a legend.

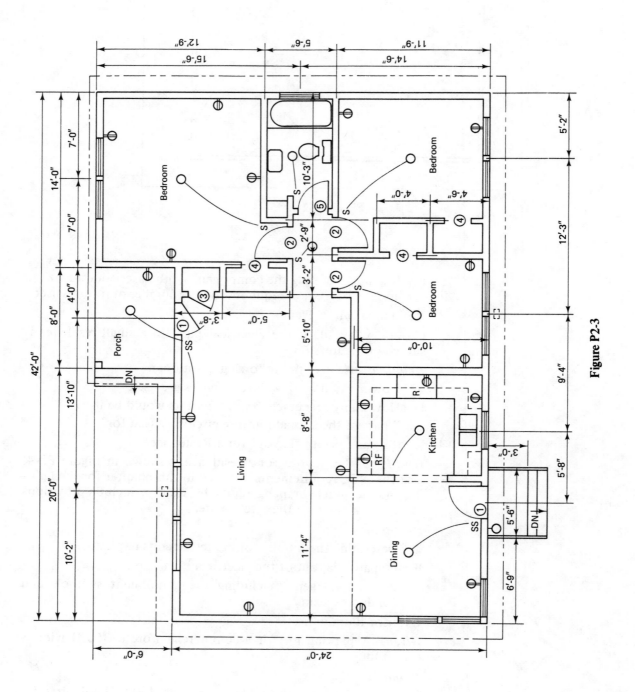

Figure P2-3

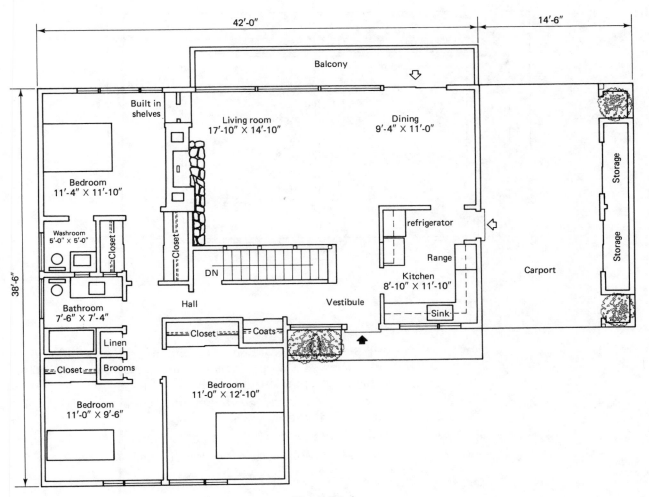

Figure P2-4

2-6 Lighting and Circuit Layouts for a Mall Clothing Store.

Given: Use the following power requirement specifications for Figure P2-6.

1. Calculate the lighting requirements for the showroom, change area, storeroom, and rest room. Use 30-fc illumination level for the storeroom and rest room. Use 60 fc for the showroom and change area. Use 40-W 3100-lm fluorescent bulbs. Each fixture contains two bulbs (except in the rest room). Use 0.65 each for the coefficient of utilization and maintenance factor. Determine the number of branch fuse circuits. Neglect lighting for special effects such as window highlighting, special display areas, and so on.

2. Calculate the duplex receptacle and special-purpose receptacle requirements. Duplex receptacles are placed every 12 ft along wall space and within 6 ft of doorways. Provisions must be made for a roof type air conditioner that consumes 25 A. Determine the wire gage and the number of branch fuse circuits.

Required: Include the following information on the layout:

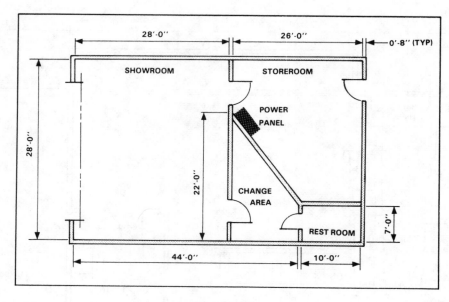

Figure P2-6

   a. All the lighting, switches, and circuits, showing the branch runs. Draw the walls, windows, doorways, and door swings using phantom and/or thin lines.

   b. Legend with component symbols, conductor sizing, and actual illumination levels.

2-7 Commercial Power and Lighting Layout.

Given: Use the following power requirement specifications for each store shown in Figure P2-7.

1. Beauty salon

   a. Overhead lighting according to the recommendations in Appendix Table 4; use standard receptacles.

   b. Air conditioner at 240 V with a compressor draw of 20.2 A and a condenser/evaporator draw of 6.4 A.

   c. Electric hot water heater rated at 5000 W and 240 V.

   d. Four hair dryers rated at 1000 W and 120 V each.

2. Pizzeria

   a. Overhead lighting according to the recommendations in Appendix Table 4; use standard receptacles.

   b. Bake oven rated load of 16,000 W and 240 V.

   c. Doughnut machine with a 2000-W heater and a 2.2-A motor at 240 V.

   d. Dough mixer rated at 15 A and 240 V.

Required: Include the following information on the layout:

a. Complete lighting-receptacle layout, broken into individual branch circuits. Draw the building walls lightly or in phantom lines.

b. Legend, including component symbols and conductor sizing.

c. Actual illumination levels used.

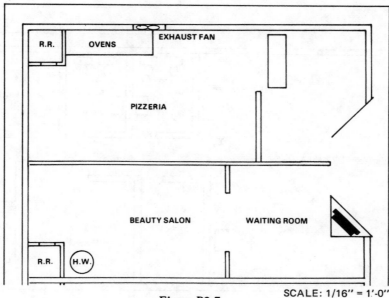

RR OVENS EXHAUST FAN

PIZZERIA

BEAUTY SALON WAITING ROOM

R.R. (H.W.)

SCALE: 1/16'' = 1'-0''

Figure P2-7

2-8 Design and Drafting Office—Power and Lighting Layout.

Given: Use the following power requirement specifications for the spaces in the building shown in Figure P2-8. The operating voltage is 120 V unless otherwise specified:

1. Design/drafting area
    a. 3-hp central air-conditioning (A/C) unit operating at 240 V (place the special-purpose receptacle in the storeroom).
    b. Overhead lighting at 100 fc (fluorescent).
    c. Standard duplex wall receptacles at 12 ft spacing and 6 ft from openings.
2. Office and conference room
    a. Overhead lighting at 80 fc.
    b. Wall receptacles (12 ft spacing).
3. Secretary/reception
    a. Special-purpose branch circuit for a copy machine operating at 1750 W.
    b. Overhead lighting at 60 fc.
    c. Wall receptacles (12 ft spacing).
4. Storage (mechanical) room and hallway
    a. 5000-W electric water heater at 240 V.
    b. Overhead lights at 40 fc.
    c. Wall receptacles (12 ft spacing).
5. Computer-aided design (CAD) room
    a. Cathode-ray-tube terminal requiring 1200 W each for two stations.

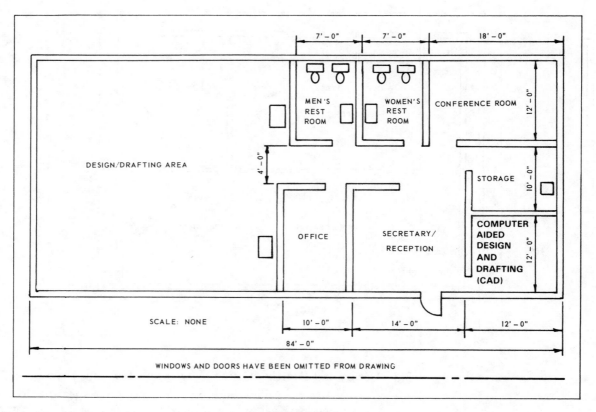

Figure P2-8

b. Pen plotter and hard-copy print device requiring 1600 W total.

c. Overhead lights at 60 fc.

d. Wall receptacles (12 ft spacing).

6. Rest rooms

a. 30-fc lighting.

b. Wall receptacles (12 ft spacing).

Required: Include the following information on the layout:

a. Complete power and lighting layout.

b. Legend, including component symbols and conductor sizing.

2-9 Recording Studio—Lighting and Circuit Layout.

Given: Operating voltage is 120 V with the exception of the special-purpose receptacles, which are 240 V. See Figure P2-9.

1. Instrument rooms

a. Each is to be illuminated with multiple incandescent lighting fixtures using dimmer switches. The illumination level is to range from 0 to 60 fc.

b. Standard duplex receptacles will be located 6 ft from doorways and 12 ft apart along the walls.

c. A special-purpose receptacle is required against the out-

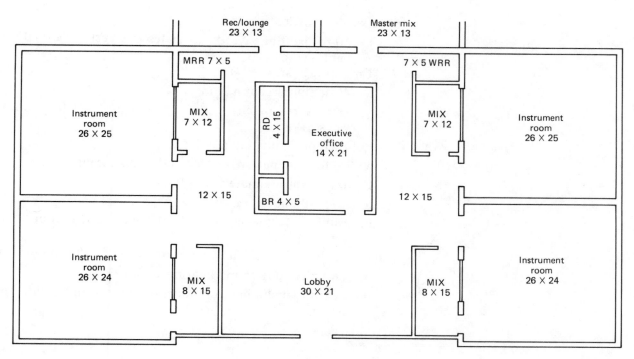

Figure P2-9

side wall of each room to accommodate a 2-hp air conditioner.

2. Mix rooms
   a. 40-fc illumination is required.
   b. Use standard duplex receptacle spacing. A floor-mounted receptacle strip containing 10 pairs of duplex receptacles is required and must be located adjacent to the viewing wall.

3. Recreation/lounge area
   a. To be illuminated with double 40-W fluorescent fixtures at a level of 60 fc.
   b. Duplex receptacles are to be spaced 12 ft apart but no farther than 6 ft from doorways.
   c. Special-purpose receptacles must be available to accommodate a 10-kW stove, a 1-hp air conditioner, and a 5000-W hot water heater.

4. Master mix room
   a. Required fluorescent illumination is 60 fc.
   b. Duplex receptacles with 12-ft spacing, 6 ft from doorways.
   c. 1-hp air conditioner.

5. Rest rooms (MRR, WRR)
   a. 60-fc lighting.
   b. Standard duplex receptacles.

6. Executive office
    a. 60-fc lighting. *Note:* Incandescent rest room (BR) lighting included.
7. Tape storage area (RD)
    a. 40-fc fluorescent lighting.
    b. Duplex receptacles with 12-ft spacing, 6 ft from doorways.
8. Lobby (approximately 630 ft^2.)
    a. 40-fc illumination, double 40-W fluorescent fixtures.
9. Hallways (approximately 348 ft^2.)
    a. 40-fc illumination.
    b. Duplex receptacles with 12-ft spacing, 6 ft from doorways.

Required: Include the following information on the layout:
    a. Complete power and lighting layout.
    b. Legend, including component symbols and conductor sizing.

Power Panel and Service Requirements

3-1 POWER PANEL

Each branch circuit is wired to a fuse or a circuit breaker. If, for any reason, the current exceeds the safe carrying capacity of the conductor, the fuse will blow or the breaker will trip, cutting power to that circuit. The overcurrent protection devices are normally placed in a power panel. The power panel is an enclosure that houses the fuseholders or circuit breakers. Each branch circuit, or home run, as it is also called, is connected to one fuse or breaker in the panel. Each overcurrent device trip rating is based on the ampacity of that particular branch-circuit conductor. A typical power panel is illustrated in Figure 3-1.

A layout of a power panel is normally prepared by a design drafter. After each branch circuit has been designed, the overprotection devices are sized accordingly. The result is illustrated as a power panel on an engineering drawing. It is used to accompany the electrical plan drawing. An example of a power panel for the electrical layout of Figure 2-24 is shown in Figure 3-2(a). For larger power distribution systems such as the one discussed in Section 2-9, more than a single panel may be used. Each will be designated differently, such as power panel A, power panel B, and so on. The standard symbol used to represent a power panel is illustrated in Figure 3-2(b). Refer back to Figures 2-24 and 2-25 to locate the panels used for those projects.

Figure 3-1 Power panel

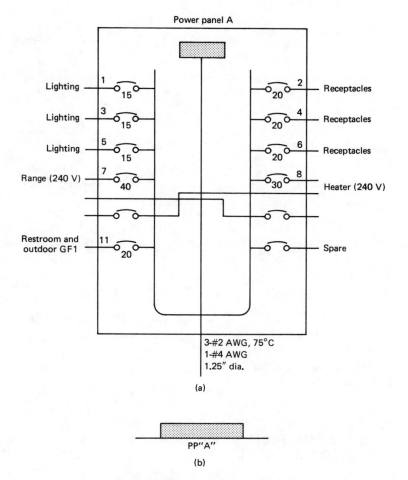

Figure 3-2   Power panel layout

Branch circuits 1, 3, and 5 of Figure 2-24 are No. 14 AWG lighting circuits. The overcurrent protection device for this purpose is 15 A. This is illustrated in Figure 3-2(a). Note also that 15 A is the minimum size allowed. The branch circuit along the upper left portion of panel A is used to represent one of the lighting circuits. It is circuit number 1, with overcurrent protection °15A° as noted. Its purpose is specified as lighting. On the actual piece of equipment, the function should be indicated on the panel directory. Locate it on the inside of the door of an enclosure similar to the one shown in Figure 2-8(b). The same format is used for the other branch circuits, such as receptacles 2, 4, and 6. A higher voltage, in this case 240 V, is shown differently, as illustrated by the range and heater circuits 7 and 8. Various ways may be used to illustrate this. One method is shown in Figure 3-2. An alternative is by showing double breakers, as in Figure 3-3. Also, the operating voltage level may or may not be specified at the trip-device location.

As mentioned previously, the NEC requires ground-fault interruption (GFI) for residential rest room and outdoor applications. An example of this is shown in Figure 3-2(a). The information given below

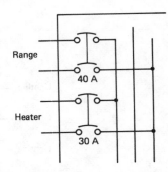

Figure 3-3   240-Volt branch circuit

the power panel refers to the conductor and conduit sizes for the incoming service. This is read as:

- 3–#2 AWG, 75°C   Three No. 2 AWG current-carrying conductors based on a 75° temperature rise
- 1–#4 AWG   One No. 4 AWG neutral ground
- 1.25″∅   One 1.25-inch-diameter conduit that carries the conductors

The origin of this information will be studied later in this chapter.

3-2 COMMERCIAL POWER PANEL

Figure 3-4 shows the power panels used for the service store commercial electrical plan in Figure 2-25. Figure 3-4(a) represents power panel A for Figure 2-25. This is used for the lighting and receptacle branch circuits. Note that additional information has been given. This is the column marked "load" next to the overcurrent device trip size marked "breaker." It specifies the design value for line current. The smallest (and only) overcurrent-device trip rating is given as 20 A.

Figure 3-4(b) represents power panel P for Figure 2-25. This is used for each of the special-purpose loads used. Both the air-conditioning and air-compressor circuits operate by using three current-carrying conductors. The balancer utilizes two. Note the way in which these are shown.

3-3 SERVICE REQUIREMENTS

The required size of a conductor leading into a power panel depends on the service requirement. The service requirement is determined by each individual branch-circuit demand. A compilation of the total demand is referred to as a *load summary*. The load summary may be determined by various methods. A common method following the NEC minimum service requirements is described below.

A single-family dwelling has an occupied floor area of 2000 ft². In addition, it has a 10-kW range.

    1. The load summary may be computed as follows:

        a. Lighting load
           2000 ft² × 3 W/ft² = 6000 W (Appendix
           Table 4)   6,000 W

LIGHTING PANEL BOARD – 'A'

Circuit number	Description	Amps.		Solid neutral	Amps.		Description	Circuit number
A-1	SALES LIGHTING	10	20		20	15	SALES RECEPTACLES	A-2
A-3	" "	10	20		20	15	SALES RECEPTACLES	A-4
A-5	" "	10	20		20	5	RESTROOM RECEPTS	A-6
A-7	TRUCK TIRE LIGHTING	6	20		20	4	RESTROOM LIGHT & FAN	A-8
A-9	FLOODLIGHTING	5	20		20	5	OFFICE	A-10
A-11	STORAGE LIGHTING	9	20		20	15	TRUCK RECEPTACLES	A-12
A-13	" "	8	20		20	15	EXTERIOR RECEPTACLES	A-14
A-15	SPARE		20		20		SPARE	A-16
		LOAD	BREAKER		BREAKER	LOAD		

(a)

POWER PANEL BOARD – 'P'

Circuit number	Description	Amps.		Solid neutral	Amps.		Description	Circuit number
P-1	AIR CONDITION UNIT	63	100		20	14	AIR COMPRESSOR	P-2
P-3	" " "	"	"		"	"	" "	P-4
P-5	" " "	"	"		"	"	" "	P-6
P-7	UNIT HEATER	5	20		20	5	UNIT HEATER	P-8
P-9	BRAKE DRUM LATH	11	20		20	15	HOIST CONSOLE	P-10
P-11	UNIT HEATER	5	20		20	15	WHEEL ALIGNER	P-12
P-13	UNIT HEATER	5	20		20	15	BENCH GRINDER	P-14
P-15	BALANCER	20	30		20	15	SINGLE RECPT	P-16
P-17	"	20	30		20	15	SINGLE RECPT	P-18
P-19	SPARE	15	20		20	15	SINGLE RECPT	P-20
		LOAD	BREAKER		LOAD	BREAKER	Size breakers and cable for power panel board per latest *National Electrical Code*	

(b)

Figure 3-4 Commercial power panels: (a) lighting and receptacle; (b) special

b. Receptacle load
two 20-A small appliance loads and one
20-A laundry load each rated at 1500 W
(Section 2-4)   4,500 W

                                                                                                          10,500 W

The lighting and receptacle load may be computed based on a demand factor according to
Appendix Table 14 as follows:
first 3000 W at 100% =   3,000 W
from 3001 W at 35% = 10,500 − 3000 =
7,500 W at 35% =   2,525 W

   c. Special-purpose load
Range: the range load (Figure 2-23,
column A) =   8,000 W

Total load = 13,525 W
= 13.525 kW

Note: The demand factor used from Appendix Table 14 is for feeders only (sum of branch loads) and not individual branch circuits.

2. Next, the service requirement may be determined. Since voltage normally will be supplied to a residence through a three-wire feeder at 240 V, the service will be based on this value.

$$\text{service} = I \text{ (current)} = \frac{P \text{ (power)}}{E \text{ (EMF)}}$$

$$I = \frac{13{,}525}{240} = 57 \text{ A (minimum)}$$

Since the net load exceeds 10 kW, however, service conductors shall be increased to 100 A for a residential application per NEC minimum requirements.

3. The required conductor size may be determined. Appendix Table 15 lists the minimum-size conductor permitted for residential service.

    For 100-A service this is No. 4 AWG copper three-wire single-phase conductor feeder. A variation of this selection may be found in Appendix Table 3(a).

    A reduced size for the neutral wire in the feeder is permitted. This may be calculated, however, but is usually two trade sizes smaller (one to be safe). Thus No. 6 AWG may be used.

    The power panel for this example is shown in Figure 3-5. Four 15-A lighting branch circuits have been used. This is based on

$$\text{lighting load} = I \text{ (current)} = \frac{P \text{ (power)}}{E \text{ (EMF)}}$$

$$I = \frac{6000 \text{ W}}{120 \text{ V}} = 50 \text{ A}$$

As mentioned, optional residential calculations may be used. One method calculates the heating and air-conditioning loads and adds the total of the first 10 kW plus 40% above 10 kW of other loads (lighting, receptacles, and so on). Others used for multifamily dwellings and schools, are given in NEC Article 220, Sections 30 through 35. The method outlined above (step 1) may be used for all residential projects in this text.

The service requirements for a commercial installation may also be determined. If the electrical layout presented in Figures 2-24 and 3-2 is for a store, the service requirements may be determined as follows:

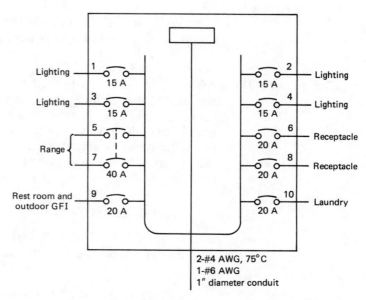

Lighting — 1 15 A
Lighting — 3 15 A
Range { 5 / 7 40 A
Rest room and outdoor GFI — 9 20 A

2 15 A — Lighting
4 15 A — Lighting
6 20 A — Receptacle
8 20 A — Receptacle
10 20 A — Laundry

2-#4 AWG, 75°C
1-#6 AWG
1″ diameter conduit

Figure 3-5 Single-phase residential power panel

1. Load summary
   a. General lighting
      1392 ft² × 3 W/ft² × 1.25 =   5,220 W
      (1.25 is a 125% derating factor
      used for continuous load)
   b. Receptacle load
      20 receptacles × 180 VA =   3,600 W
      (must use not less than 180 VA to
      figure each duplex receptacle load)
      The entire receptacle load is used
      since it is less than 10 kW. Use 50%
      of all above 10 kW for a commer-
      cial application.
      _____
                                                           8,820 W
   c. Special-purpose loads
      Range   7,000 W
      Electric water heater   5,000 W
      _____
                             Total load = 20,820 W
                                        = 20.82 kW

2. Service requirement

$$I = \frac{20{,}820 \text{ W}}{240 \text{ V}} = 87 \text{ A}$$

3. Minimum feeder size
   The service conductors would have to be at least No. 4 AWG, 75°C. In Figure 3-2, a three-wire No. 2 AWG, 75°C, with a No. 4 AWG ground was used. This will safely carry over 100 A at 240 V regardless of the type of insulation used.

Note: If a show window had been used in the store, an additional lighting load of 200 W per linear foot would have to be added.

3-4 CONDUIT

After branch-circuit and feeder conductors have been sized, the manner in which the run is made should be determined. For many applications, especially hazardous areas, it is desirable, and sometimes mandatory, to provide protective covering for the conductors. A common method employs conduits. The conductors are carried inside a conduit such as the one shown in Figure 3-6. The required conduit size will depend on:

1. The AWG size of conductors
2. The number of conductors
3. The type of conductor insulation

Appendix Table 16 designates these variables and indicates the required conduit size for combinations. For example, a 1-in.-diameter conduit will carry up to 25 No. 14 AWG conductors having type XHHW insulation. If 26 conductors are used, a 1.25-in. diameter is required. Note that the minimum conduit size shown in the table is 0.50 in., the commercial minimum that may be used.

Different types of insulation are available; each is utilized for a specific application. Depending on the type used, the thickness, and thus the conduit diameter, will vary. Common types of insulation likely to be encountered are the following:

type letters	trade name
RUH	Heat-resistant latex rubber
RHW	Moisture- and heat-resistant rubber
RUW	Moisture-resistant latex rubber
THHN	Heat-resistant thermoplastic
THW	Moisture- and heat-resistant thermoplastic
THWN	Moisture- and heat-resistant thermoplastic
XHHW	Moisture- and heat-resistant cross-linked synthetic polymer
T	Thermoplastic
UF	Underground feeder and branch circuit

Refer to NEC Article 310 for special insulation application conditions. The feeder into the power panel shown in Figure 3-2(a) uses

Figure 3-6 Conduit (electrical metallic tubing)

three No. 2 AWG and one No. 4 AWG conductors. For simplicity in sizing the required conduit, assume all four conductors to be No. 2 AWG. This can be done since the No. 4 is actually smaller in diameter and uses less space. Referring to Appendix Table 16, we find that a 1.25-in.-diameter conduit will safely carry four No. 2 AWG with type RUH insulation. Thus the feeder conduit used is 1.25 in. in diameter, as shown in Figure 3-2(a). The conduit for the feeder conductors used in Figure 3-5 is sized in a similar manner. Two No. 4 AWG and one No. 6 AWG conductors are used. The NEC allows a 1-in.-diameter conduit to carry three No. 4 AWG conductors of type T, RUH, or RUW insulation.

An application may require carrying conductors that vary considerably in size within the same conduit. This is permissible as long as the operating voltage levels are within the same range. All conductors under 600 V may be carried in the same conduit. Residential and commercial applications normally fall within this voltage range. Conductors operating at more than 600 V cannot occupy an enclosure with those operating at less than 600 V. To determine the minimum conduit size required, the percent of "area fill" must be considered. If there are two or more conductors in a conduit, the maximum fill is 40%. This may seem low; however, the length of pull and number of bends must be considered. Normally, 360° of bend (four 90° bends) is the maximum allowed per conduit run. Also, the conduit bend radius must be generous. Appendix Table 17 lists minimum required bend radii. If the length of run and bend angle are too great, it is possible to exceed the conductor tensile strength as listed in Appendix Table 12. Thus the 40% maximum fill shall be used for all projects.

To determine the required conduit size for various conductors, use Appendix Tables 18 and 19. First determine the total area of the conductors from Appendix Table 19. For example, if three No. 3 AWG and one No. 4 AWG type RUH conductors are to be carried in a single conduit, the total conductor area is

$$\text{total conductor area} = 3 \times \text{No. 2 AWG area} + 1 \times \text{No. 4 AWG area}$$
$$= 3 \times 0.1473 \text{ in.}^2 + 1 \times 0.1087 \text{ in.}^2$$
$$= 0.5506 \text{ in.}^2$$

Next, refer to Appendix Table 18. For more than two conductors that are not lead covered, a 1.25-in.-diameter conduit allows a fill area of 0.60 in.^{2}. A 1.00-in. diameter allows only 0.34 in.^{2}. Thus a 1.25-in.-diameter conduit is selected since this is the next largest size above the minimum conductor area. This result corresponds to the earlier approximation method used for the example problem shown in Figure 3-2(a). The example above uses rigid metal conduit, which is permissible for use in all types of atmospheric conditions.

Some conductor runs will involve underground applications. It may be permissible for many of these applications to use direct burial cable, which is shielded or sheathed. Thus conduit or tubing is not needed, as the cable will be run directly in a trench that has been dug to the correct depth. For these applications, use type UF (underground feeder and branch circuit) cable. Include this information directly on the power and distribution layout in the form of a note.

Figure 3-7   Cable tray

3-5 CABLE TRAYS Commercial and industrial applications often use a cable tray to carry the conductors. A cable tray is an enclosure in which conductors are set per the requirements of NEC Article 318. The enclosure may be one of several styles: ladders, troughs, channels, or solid-bottom types. A typical ladder/ventilated trough type of tray is shown in Figure 3-7. They may have open or covered tops, depending on the application. Since they are usually larger than conduits, more wires can be accommodated. The general requirements for multiconductors in ladder or ventilated troughs are as follows:

1. *All cables smaller than No. 0000 AWG:* The sum of the cross-sectional areas of all the insulated cable (as determined from Appendix Table 19) shall not exceed the maximum cable fill area of Appendix Table 20, column 1.

2. *All cables larger than No. 0000 AWG:* The sum of the cable diameters shall not exceed the inside tray width. The cables are installed in a single layer.

3. *Trays containing cables both larger and smaller than No. 0000 AWG:* The sum of the cross-sectional areas of the cables less than No. 0000 AWG shall not exceed the maximum fill area of Appendix Table 20, column 2. The cables greater than No. 0000 AWG are to be installed in a single layer with no other conductors over them.

Note: For solid-bottom cable trays, refer to Appendix Table 20, columns 3 and 4, for maximum fill area.

For example, a 6-in.-wide × 4-in.-high cable tray is used on a commercial installation. Referring to Appendix Table 20, column 1, a maximum fill area of 7 in.^{2} (6.50 in.^{2} for single conductors) may be used for cables smaller than No. 0000 AWG. Next, determine the cross-sectional area of the conductors from Appendix Table 19. No. 6 AWG conductors with type XHHW insulation are to be carried in the tray. Each has a cross-sectional area of 0.0625 in.^{2}. Thus

$$\frac{6.50 \text{ in.}^2}{0.0625 \text{ in.}^2/\text{conductor}} = 104 \text{ single conductors}$$

may be carried.

Cable trays may be attached to structural supports in a manner similar to that shown in Figure 3-8. They are held up by square tubing and threaded rod assemblies. One method used with ladder or open-bottom trays is to tie the cables approximately every 3 ft, as illustrated. Standard fittings are available for cable tray installations. These include such items as 90° left and right elbow, 45° left and right elbows, tee, cross, vertical drop, and vertical lift. An installation using several of these fittings is shown in Figure 3-9. In general, layout technique involves using as much straight run as possible and avoiding the use of fittings, since they are costly.

3-6 SERVICE DIAGRAM

After the power and distribution for a commercial electrical application has been designed and the layout is complete, the service requirements are specified. For larger systems consisting of more than one power panel, this may be accomplished by using a diagram. The type of schematic representation shown in Figure 3-10 is often referred to as a *riser diagram*. It shows all panels, each meter with secondary fuses, main switch, voltage step-down transformer to supply, and ground. Power panels A and B in Figure 3-10 may be either part of a single occupancy or two separate occupancies, depending on the type of application. Panels A and B, for example, may be used for different stores in a mall. A separate electric meter is installed for each. The conductors, conduits, and fuses have been sized per the branch-circuit feeder requirements developed in Chapter 2. The secondary fuses are rated at 125% of full-load current. The transformer and grounding may or may not be shown on any particular diagram.

An example of a riser diagram for the commercial power panel layouts in Figure 3-4 is shown in Figure 3-11. Note the differences between this and the preceding diagram. The information may be presented in several ways. Methods will vary somewhat from company to company. The conductor, conduit, and fuse sizes, however, will normally be shown in each case.

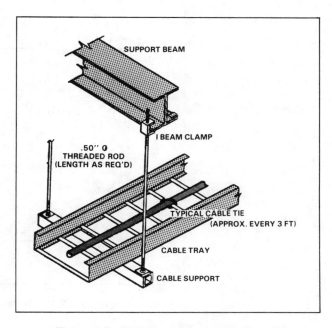

Figure 3-8 Cable tray hanger installation

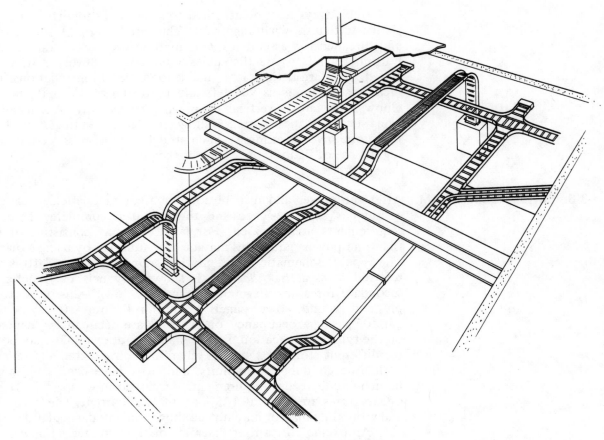

Figure 3-9 Cable tray system

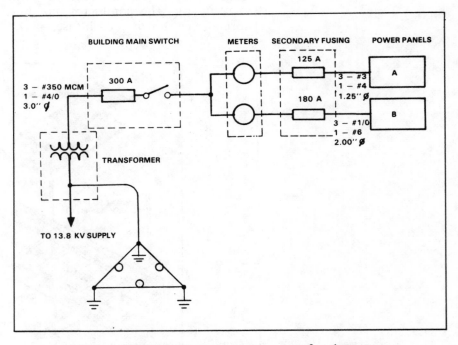

Figure 3-10 Service entrance requirements for plaza stores

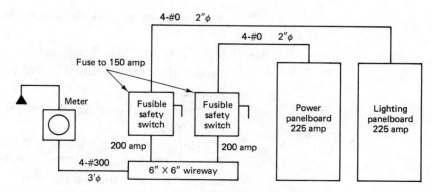

Figure 3-11 Commercial service

ASSIGNMENTS

Project work in Chapter 3 should be accomplished with the NEC.

3-1 Room Power Requirement.

Given: This assignment is a continuation of Assignment 2-1.

Required: Determine the power requirement for the lighting and receptacle circuits of Assignment 2-1.

3-2 Office Power Requirement.

Given: This assignment is a continuation of Assignment 2-2.

Required: Determine the power requirement for the overhead-lighting layout of Assignment 2-2.

3-3 Residential Service Requirements.

Given: This assignment is a continuation of Assignment 2-3 using a 10-kW range.

Required:

a. Compute the total power requirements for lighting, receptacle, and range loads.

b. Specify the minimum service requirements using the method described in Section 3-3. Include the feeder and conduit sizes required.

3-4 Residential Power Panel and Service.

Given: This assignment is a continuation of Assignment 2-4.

Required:

a. Prepare a power panel to meet the requirements of Assignment 2-4.

b. Compute the total power requirements for lighting, receptacle, and range loads.

c. Specify the minimum service requirements using the method described in Section 3-3. Include the feeder and conduit sizes required.

3-5 Cable Tray.

Given: Fifteen No. 00 AWG type RHW insulation and five No. 1 AWG type RHW insulation single conductors are to be carried by an open-type cable tray.

Required: Determine the minimum-size cable tray that may be used.

3-6   Mall Clothing Store Power Panel and Service.

Given: This assignment is a continuation of Assignment 2-6.

Required:

a. Prepare a power panel to meet the requirements of Assignment 2-6.

b. Compute the total power requirements for lighting, receptacle, and special-purpose loads.

c. Specify the minimum service requirements, including the feeder and conduit sizes required.

d. Prepare a riser diagram.

3-7   Commercial Power Panel and Service.

Given: This assignment is a continuation of Assignment 2-7.

Required:

a. Prepare two power panels to meet the requirements of Assignment 2-7. One panel is for the pizzeria and the other for the beauty salon.

b. Compute the total power requirements for lighting, receptacle, and special-purpose loads.

c. Specify the minimum service requirements, including the feeder and conduit sizes required.

d. Prepare one riser diagram which includes both power panels.

3-8   Design and Drafting Office—Power Panel and Service.

Given: This assignment is a continuation of Assignment 2-8.

Required:

a. Prepare a power panel to meet the requirements of Assignment 2-8.

b. Compute the total power requirements for lighting, receptacle, and special-purpose loads.

c. Specify the minimum service requirements, including the feeder and conduit sizes required.

d. Prepare a riser diagram.

4

Special Occupancy

4-1 HAZARD CATEGORIES

A variety of situations fall under the category of "special occupancy," which will be covered in this chapter. The foremost special situation involves a hazardous location. A hazardous location exists whenever sparking could result in an explosion or fire, thus threatening life and property. Hazardous locations fall into three categories, including:

1. *Class I:* Class I encompasses areas containing flammable gases or vapors, including paint spray booths, gasoline dispensing pumps, refineries, and chemical plants.
2. *Class II:* An area containing combustible dust is a class II location. Examples include coal pulverizing and grain or feed processing plants.
3. *Class III:* An area is considered class III if ignitable fibers are handled. An example of this would be a clothing manufacturer.

Hazardous location applications thus can be seen to be industrial as well as commercial. To identify further the type of hazard and its application, the materials are grouped in A, B, C, D, E, F, and G categories. Groups A, B, C, and D fall under class I. The group specification depends on the exact chemical(s) in the location. Refer to *National Electrical Code®* Table 500-2 for a material listing. In general, however, rating is based on ignition temperature. Groups A, B, and D have ignition temperatures of 536°F and group C is 356°F. All equipment must be approved and marked for the class and group number. The maximum operating temperature of the component must also be marked. Special enclosures are used for hazardous locations. The National Electrical Manufacturers Association (NEMA) has standardized enclosures for this purpose. A class I, group C or D enclosure, for example, is a NEMA type 7. It is designed for use in hazardous atmospheres that contain petroleum, gasoline, alcohol, acetone, or lacquer vapors. The enclosures are made of cast gray iron and machined to provide a smooth metal-to-metal seat between the enclosure cover and body. A NEMA type 7 enclosure is shown in Figure 4-1.

Figure 4-1 NEMA 7 hazardous location enclosure (Courtesy of Allen Bradley)

Groups E, F, and G fall within class II. The groups are categorized by the type of combustible dust material in the location. Group E is for metal (e.g., aluminum) dust. Group F is a nonmetal dust category. Materials within this group include coke and coal. Group G also includes nonmetal dust created during grain and flour operations. NEMA type 9 hazardous dust enclosures may be used for these purposes. As with NEMA type 7, they provide a smooth metal-to-metal seat between the enclosure cover and body. A NEMA type 9 enclosure is shown in Figure 4-2. If the dust material is not combustible, a NEMA type 12 dust-tight enclosure may be used. They are constructed of steel sheet metal with gaskets. The gaskets prevent dust from entering and accumulating on the component(s) switch contacts. A NEMA type 12 enclosure is shown in Figure 4-3. As with class I, class II equipment must be approved and marked with the appropriate NEMA number. Class III has no group designations.

Besides class and group of combustible materials, there is also placement in a division. These include the following:

1. *Class I, division1:* for situations when gases and vapors are normally present.
2. *Class I, division 2:* for situations where gases and liquids are kept in containers.
3. *Class II, division 1:* for situations when combustible dust is normally present.
4. *Class II, division 2:* for situations when dust is not present in the air; however, particles may accumulate on the components.

Figure 4-2 NEMA 9 hazardous dust enclosure (Courtesy of Allen Bradley)

Figure 4-3 NEMA 12 dust-tight enclosure (Courtesy of Allen Bradley)

5. *Class III, division 1:* for situations involving the use and manufacture of fiber.
6. *Class III, division 2:* for the storage of fiber products.

4-2 CLASS I LOCATIONS

The rules of NEC Article 501 must be followed for all class I locations. Explosion-proof equipment requirements are outlined in the Article. Use a NEMA type 7 or 9 enclosure for each component, or to cover groups of components on control panels.

Wiring methods vary depending on the division. Division 1 allows three methods for conductor installation. These include:

1. Rigid metal conduit
2. Mineral-insulated (type MI) cable
3. Threaded steel intermediate conduit

Only threaded connections are allowed with explosion-proof boxes and fittings, as shown in Figure 4-4.

Division 2 allows additional wiring methods, including the three listed under division 1, plus the following:

1. Enclosed gasketed busways
2. Metal-clad (type MC) power cable
3. Shielded nonmetallic sheathed (type SNM) cable
4. Power and control tray (type TC) cable
5. Medium-voltage (type MV) cable
6. Remote control signaling circuit (type PLTC) cable

Boxes and fittings do not have to be explosion proof. If a seal is required, however, the seal fitting must be explosion proof.

Seals are used for hazardous-location installations. When conduit enters a piece of equipment, it is sealed near the equipment, usually within 18 in. This isolates that component from other parts of the wiring system. Any explosion in that piece of equipment thus cannot

Figure 4-4 Box and fitting threaded connections (Courtesy of General Electric Corp.)

travel along the conduit since it is blocked (sealed). Seals must be provided in the conduit where it enters circuit breakers, fuses, switches, relays, resistors, and other devices that could produce arcing, sparking, or high temperature. This includes both divisions 1 and 2 except for general-purpose enclosures. Seals are also required when conduit leaves a hazardous area. This prevents hazardous atmosphere from traveling back to a "safe" area such as a control room.

Class I requires special provisions for lighting fixtures and grounding. Suspended-type lighting, for example, must be wired by metal conduit. Also, all metal equipment must be grounded. Again, refer to Article 501 for specifics.

A commercial application such as an anesthesia room is considered a hazardous condition. This requires a special branch circuit because of the explosion-proof nature of the room. The branch circuit must be run in a conduit. First, the size of the conductors is determined by the branch-circuit load requirements. Refer to Chapter 2 for detailed methods. Next, determine the number of conductors to be carried. Then select the minimum size for the rigid metal trade conduit from Appendix Table 16. Heat generation by the conductors may create a problem. Base the conductor size on the 75°C (167°F) temperature-rise column in Appendix Table 16 [98°C (185°F) is the maximum allowable]. Also, the greater the number of conductors carried in a conduit, the greater the heat generation. Thus limit each run to three conductors that carry current. These measures will assure safety compliance.

4-3 CLASS II AND III LOCATIONS

The rules of NEC Article 502 must be followed for all class II applications. The equipment used must either withstand "dust igniting" without transmitting an arc or spark (NEMA type 9) or be dust tight (NEMA type 12). Other equipment requirements are outlined in the Article.

Class II, division 1 locations require that the wiring method use either (1) metal conduit or (2) mineral-insulated (type MI) cable. Division 2 allows, in addition to the conduit and cable listed above, the following:

1. Dust-tight wireways
2. Metal-clad power (type MC) cable
3. Intermediate metal conduit
4. Shielded nonmetallic sheathed (type SNM) cable
5. Electrical metallic tubing (EMT)

As in class I locations, seals are required for class II. Long distances between components may be exempt from seals, however, since dust will not travel over 5 ft upward or 10 ft horizontally. Additional consideration must be made for surge protection and lighting fixtures. Again, refer to Article 502 for specifics.

The surface temperature of operating equipment must be maintained within certain limits for class III locations. Maximum temperature for components not subject to overload is 329°F. For components subjected to overload, the maximum limit is 248°F. U.L.-approved equipment meets these requirements.

Wiring methods for divisions 1 and 2 generally allow:

1. Threaded rigid metal conduit
2. Threaded steel intermediate metal conduit
3. Type MI cable
4. Type MC cable

As with classes I and II, specific requirements are made for all equipment, lighting fixtures, receptacles, and grounding. Refer to Article 503 for specifics.

Many industrial applications now use "intrinsically safe" methods. Intrinsic safety involves making electrical equipment safe for hazardous locations by limiting the available energy. The level will be too low for it to create a hot enough spark, or heat a surface hot enough, to ignite a gas, vapor, or dust. This will vary with the specific environment. The use of intrinsically safe equipment also increases the safety of personnel. For example, some companies limit control operating voltage to 24 V in any area where an operator might be exposed to the voltage. NEC Articles do not apply in low voltage applications.

The advantage of intrinsically safe equipment is that an accident involving wiring cannot cause an explosion. Such equipment is foolproof, thus eliminating the need to enclose the conductors in rigid conduit with explosion-proof fittings. The disadvantage is that the low energy limits the use of such equipment to lower-power devices. It cannot be used to operate large motors or general lighting.

4-4 COMMERCIAL GARAGES AND SERVICE STATIONS

Wherever auto repair, service work, and "volatile flammable liquid" dispensing is done, the electrical installation will conform to NEC Articles 511 and 514. The general features include:

1. Marking. All equipment marked with "approved for class I locations." Refer to Section 4-1 for class I requirements.
2. Hazard areas
   a. Garage area (shown in Figure 4-5)
      • Class I, division 1: Any pit below floor level.
      • Class I, division 2: The entire servicing floor area up

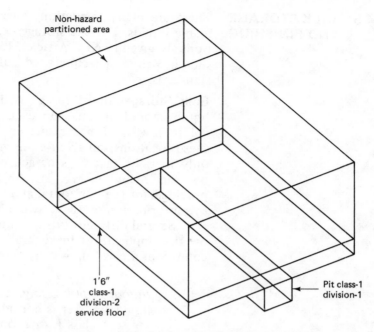

Non-hazard
partitioned area

1'6"
class-1
division-2
service floor

Pit class-1
division-1

Figure 4-5 Commercial garage hazard area

to 18 in. above the floor. Within the 18-in. imaginary plane above the floor:

1. Conductors are in conduit.
2. Must specify seals for conduit greater than 2 in. in diameter if broken by junction boxes or couplings.
3. Specify seals for entry into switch, breaker, fuse, relay, and resistor/enclosures.

Above the 18-in. imaginary plane above the floor:

1. Conductors are in conduit. Use either rigid metal or EMT.
2. Receptacles and plugs must be of the "polarized" type.
3. All equipment is totally enclosed.
4. Light fixtures require a lens or globe.

- Nonhazard: Adjacent partitioned (solid wall) areas.

b. Dispensing area

- Class I, division 1: Floor area within a distance of 18 in. and 4 ft up from the base. Also, an area within a distance of 3 ft away from vent pipes.
- Class I, division 2: Floor area within a distance of 20 ft away and an imaginary height of 18 in. above ground. Also, a distance 10 ft away and 18 in. above ground from a fill pipe. Also, an area from 3 to 5 ft away from vent pipes.
- Nonhazard: A solid wall partitioning within division 2 areas.

4-5 BULK STORAGE AND FINISHING

There are many regulations governing the storage and use of flammable liquids. For the storage of bulk flammable liquids, the regulations are given in NEC Article 515. For the use of flammable finishes, such as paints or lacquers, consult NEC Article 516 and other NFPA standards.

Bulk Storage Bulk storage is defined as a quantity of flammable liquid, stored in tanks, in excess of one carload (one tank truck in volume). Class I hazardous conditions exist in specified areas. In these locations, all fixed equipment must be totally enclosed. Also, only metallic raceways or acceptable types of cable insulation may be used, including M1, TC, SNM, or MC. Approved seals must be provided for the horizontal and vertical raceway boundaries. Indicate this information on the drawing layout.

Several different types of bulk storage locations exist, depending on the application. In general, the requirements for two common conditions are as follows:

1. *Indoor warehousing:* A storage area having no flammable liquid transfer is considered ordinary. If, however, an opening to a class I, division 1 or 2 area exists, the entire room must be classed the same as the adjoining area.
2. *Drum/container filling:* Adequate ventilation is required for indoor applications. It is considered division 1 within 3 ft of vent and fill openings. Between 3 and 5 ft from vent and fill openings, and up to 18 in. above the floor within a 10-ft radius, a division 2 condition exists.

Finishing A class I situation is defined with exposure to flammable vapors such as paint or lacquers; and when the finish of a part is sprayed, brushed, or dipped frequently. In general, divisions are classified by NEC as:

1. *Class I, division 1:* the interior of spray booths and exhaust ducts and any space in the direct path of spray or with air-suspended flammables
2. *Class I, division 2:* within 20 ft horizontally and 10 ft vertically of an open-spraying division 1 location

Other division 1 and 2 locations exist; however, they will not be covered in this text. Only those relating to the project at the end of the chapter have been given. Refer to NEC Article 516 if another condition is encountered. For the project in this text, determine which class I locations are division 1 or 2. Indicate this on the layout, including the necessary component enclosures and conduit for a hazardous location.

4-6 PLACE OF ASSEMBLY

A room is considered a place of assembly when a minimum of 100 persons gather there. This category includes such buildings as halls, restaurants (dining), auditoriums, and various municipal/commercial building areas. The dining area for the project at the end of the chapter will be considered a place of assembly. Use the following wiring methods:

1. Metallic or nonmetallic raceways encased in at least 2 in. of concrete
2. Mineral insulated, metal sheathed, or type MC cable

Specify the raceways and cable to be used on the layout. Also specify that the use of exit lights is required.

4-7 HEALTH CARE FACILITY

The NEC has a lengthy Article for applications involving health care facilities. The general requirements for systems involving flammable vapors, alternate power, and alarms are covered here. For other applications, refer to NEC Article 517 or the following:

article	purpose
700	Emergency systems
701	Standby systems
760	Fire protective signal systems
800	Alarms and communications

Flammable Vapor Any area intended for the administering of anesthetic inhalation may produce flammable vapors. This includes an anesthesia room in an extended health care facility or nursing home, and is a class I location. Design and lay out the receptacle branch circuit in this area on a separate circuit. That is, do not connect receptacles in other locations as part of the circuit. The receptacles will be of the type to accommodate attachment plugs for such locations. List these as hazardous class I, group C devices. Connect the receptacles with correctly sized conductors carried within metal conduit.

All other equipment in the area that could produce a spark shall be approved for hazardous atmospheres. Each must be totally enclosed and explosion proof, including lamps and lampholders for fixed lighting. Specify each of these requirements on the project layout.

Alternate Power An emergency system for alternate power shall be provided with a generator set or separate service as follows:

1. Power for essential loads includes illumination for safe exit (including corridors, dining, and recreation areas), exit sign(s), alarms, and elevator operation.
2. A minimum of two branches are mandatory and must be kept entirely independent of all other wiring and equipment. One branch, known as the *critical branch*, is for operation of the elevator. The other, known as the *life safety branch*, is connected to the other functions listed in (1).
3. Automatic switch equipment: locations are convenient to authorized personnel.

4. Overcurrent protection is accessible to authorized personnel only.

Alarms Provide an alarm system. The fire alarm may be activated at manual stations. An electric water flow alarm is connected to the sprinkler system.

For the health care project in this chapter, add a note to the layout indicating that alternate power and alarms are required.

ASSIGNMENTS

Chapter 4 projects are extended versions of Chapter 2 and 3 projects. Additional special requirements have been imposed, however. Refer to the appropriate NEC Article for each special occupancy requirement. Do not attempt to solve these projects until the concepts presented in Chapter 2 and 3 are understood.

4-1 Chlorine Room Receptacle.

Given: A room has 18 ft × 25 ft floor dimensions and one entrance. Chlorine used for swimming pool maintenance is stored in the room. *Note:* Neglect ground-fault circuit-interrupter (GFI) receptacle protection and NEMA enclosures for the prevention of water/moisture entry for this assignment.

Required: Determine the number and type of duplex wall receptacles that would be used. How many branch circuits are required?

4-2 Furnace Charging System Power and Distribution.

Given: Use the following power requirement specifications for the blast furnace plant shown in Figure P4-2. Use class II NEMA type 9 enclosures in hazardous areas.

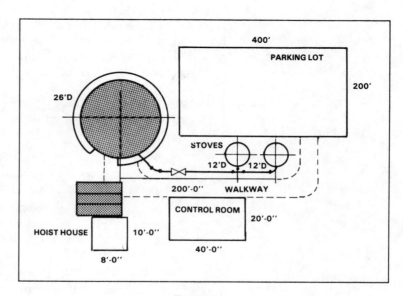

Figure P4-2

1. Hoist house
   a. Special-purpose outlet for the 40-hp motor.
   b. Light and standard duplex wall receptacle.
2. Blast furnace
   a. 30-in. circular stairway extends to the top of the 26-ft-diameter at a 45° incline, 75-ft-high furnace, illuminated at 0.50 fc in addition to one warning bulb at the top.
3. Parking lot
   a. Lighted at a 1.00-fc level.
4. Walkway
   a. Lighted at a 0.80-fc level.
5. Control room
   a. Overhead lighting at 60 fc.
   b. Standard duplex wall receptacles.
   c. Special-purpose receptacle for a 30-A control panel.

Required: Include the following information on the layout:

a. Complete power and lighting layout broken into individual branch circuits. Size and include underground conduit runs to the power panel located in the control room. Draw the background in phantom lines.
b. Legend, including component symbols and conductor sizing.
c. Power panel, including each branch-circuit breaker size and location.
d. Load summary to determine total service requirements and main supply conductor sizing.
e. Elevation of the power panel, including each branch circuit.
f. Power distribution and riser diagram from the substation to the pole (supply).

4-3 Commercial Garage Power and Distribution.

Given: Use the following power requirement specifications for the building shown in Figure P4-3.

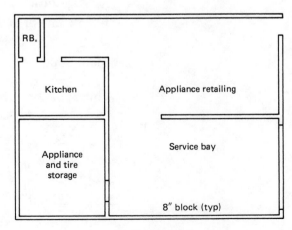

SCALE: $\frac{3}{64}$" = 1'0"

Figure P4-3

1. Retail area
   a. Manufacturing data on the air-conditioning unit: 240-V, three-phase, full-load amperage = 20 A, and 0.50-hp condenser.
   b. Overhead fluorescent lighting at 80-fc illumination.
   c. Minimum of six outlets on a special branch circuit for appliance demonstration.
   d. Standard duplex wall receptacles.
   e. Display window lighting.
2. Employee lunchroom (kitchen)
   a. Range rated at 8 kW and 240 V.
   b. Overhead incandescent lighting and standard receptacles for kitchen appliances.
3. Tire and appliance center storage
   a. Overhead lights at 40-fc illumination and standard receptacles.
4. Service bay (all eqiupment shall be specified in compliance with Section 4-4).
   a. Overhead lighting at 60-fc illumination.
   b. Receptacles for shop appliances.
   c. Hoist motor rated at 1 hp and 120 V.

Required: Include the following information on the layout:

a. Complete lighting-receptacle layout broken into individual branch circuits. Draw the building walls lightly or in phantom lines.
b. Legend, including component symbols and conductor sizing.
c. Actual illumination levels used.

4-4 Automatic Conveyor Power Distribution and Layout.

Given: Use the following power requirement specifications for the building addition shown in Figure P4-4.

1. Control room
   a. Overhead lighting level of 60 fc.
   b. Special branch circuit for a 20-A control panel.
   c. Standard duplex wall receptacles.
2. Main building

   Spray painting will be done at the end of the operation at conveyor 2 (class I, division 1). Consider the entire prepara-

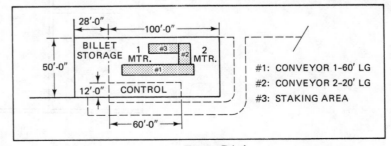

Figure P4-4

tion room to be a division 1 location. A partition isolates the control room and storage area from these vapors.

    a. Overhead lighting level of 20 fc.

    b. Special branch circuits for conveyor motors 1 and 2. Each motor is 15 hp operating at 240 V.

3. Billet storage

    a. The billet storage area is considered bulk storage.

Required: Include the following information on the layout:

a. Complete power and lighting layout broken into individual branch circuits. Size and include underground conduit runs from the parking lot and walkways to the power panel located in the control area. Size and include standard conduit runs in the building area. Draw the background in phantom lines.

b. Legend, including component symbols and conductor sizing.

c. Power panel, including each branch-circuit breaker size and location.

4-5 Chemical Processing Plant—Power Distribution and Lighting Layout.

Given: Use the following power requirement specifications for the portion of the chemical plant shown in Figure P4-5. Use class I, NEMA type 7 equipment.

1. Process equipment

    a. Special branch circuits for each of the following motors used in the process:

        (1) Main supply motor: M1 rated at 10 hp.

        (2) Secondary supply motor: M2 rated at 10 hp.

        (3) Heat exchanger motor: M3 rated at 10 hp.

2. Electrical control room

    a. Special branch circuit for the control panel rated at 40 A.

    b. Overhead lighting at 50 fc.

    c. Standard duplex wall receptacles.

3. Chlorine-processing compressor building

    a. Overhead lighting at 20 fc.

    b. Standard duplex wall receptacles.

4. Surrounding outside area

    a. Lighting at 1.00 fc up the vertical tower (gas generator) stairway, including a bulb at the top; tower elevation is approximately 45 ft due to pipe which extends vertically beyond.

    b. Lighting at the heat exchanger area.

    c. Walkway lighting between the control room and the supply motor (under the pipe rack) M2 at approximately 1.00 fc.

Required: Include the following information on the layout:

a. Complete power and lighting layout broken into individual branch circuits. Size and include conduit runs from the process equipment motors to the electrical control room.

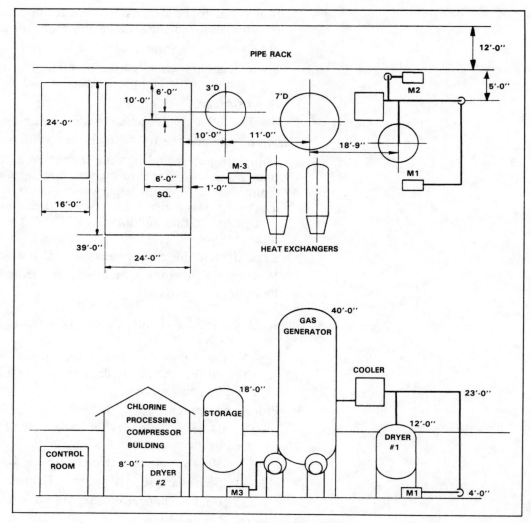

Figure P4-5

Size and include underground conduit runs where necessary. Draw the background structures lightly or in phantom.

b. Legend, including component symbols and conductor sizing.

c. Power panel, including each branch-circuit breaker size and location.

4-6 Restaurant Power and Distribution.

Given: Use the following power requirement specifications for the spaces in the building shown in Figure P4-6. The operating voltage is 120 V unless otherwise specified. Refer to NEC Article 518 for special dining-room requirements.

1. Dining room

   a. 2-hp air-conditioning (A/C) unit at 240 V. Place the special-purpose receptacle in the mechanical room.

   b. Overhead lighting at 60-fc fluorescent or 40-fc incandescent.

   c. Six floor receptacle outlets for food preparation at tables.

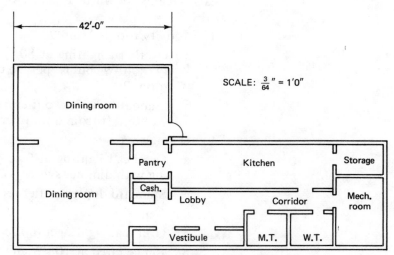

Windows and doors have been omitted from drawing

Figure P4-6

   d. Standard duplex wall receptacles spaced at 12 ft and within 6 ft from openings.

2. Kitchen/pantry

   a. Range rated at 10 kW and at 240 V.

   b. Overhead lighting at 60 fc.

   c. Heavy-duty dishwasher at 0.50 hp and 1500-W dryer.

   d. Dough mixer, which consumes 15 A.

   e. Standard duplex wall receptacles spaced at 12 ft and within 6 ft from openings.

3. Storage/mechanical rooms

   a. Overhead lights at 40 fc.

   b. 3000-W electric heater at 240 V.

   c. Wall receptacles (12 ft spacing).

4. Lobby/corridor and vestibule area

   a. Plug for electronic cash register.

   b. 30-fc lighting.

   c. Wall receptacles (12 ft spacing).

5. Rest rooms (M.T., W.T.)

   a. 30-fc lighting.

   b. Wall receptacles (12 ft spacing).

Required: Include the following information on the layout:

a. Complete power and lighting layout.

b. Legend, including component symbols and conductor sizing.

c. Compute total power requirements for lighting, receptacle, and special-purpose loads.

d. Specify minimum service requirements based on NEC demand specifications (Appendix Table 14) and 240-V supply.

4-7 Health Care Facility Power and Lighting.

Given: Use the following power requirement specifications for each area as shown in Figure P4-7. Also refer to Article 517 in NEC.

1. Activity rooms 1 and 2
    a. Overhead lighting at 80 fc using fluorescent lighting and two 40-W bulbs per fixture. Three-way switches for room 2.
    b. Standard duplex outlet receptacles (12-ft spacing and within 6 ft from door openings).

2. Dining area
    a. Overhead lighting at 0 to 40 fc using incandescent lighting with dimmer switch(s). Use the symbol S_{dim}.
    b. Standard duplex outlet receptacles (12-ft spacing).

3. Kitchen
    a. Overhead lighting at 60 fc.
    b. Range rated at 10 kW and 240 V.
    c. Waste disposal unit rated at 11.2 A and 120 V.
    d. Automatic conveyor dishwashing unit at 1.50 hp and 3000-W heater for drying.
    e. Standard duplex outlet receptacles (12-ft spacing).

4. Maintenance and boiler room
    a. Overhead lighting at 40 fc.
    b. Electric-fired heating unit at 100 A and 240 V.
    c. Standard and special receptacle power panel location.
    d. Standard duplex outlet receptacles (12-ft spacing).

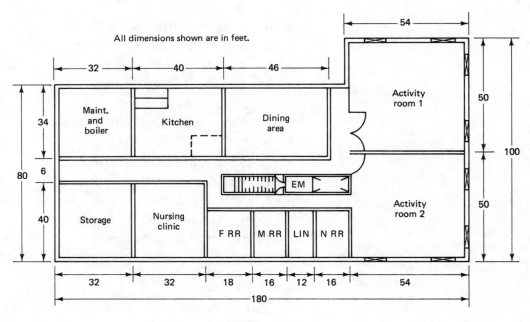

Niagara County Health Care Facility
(first floor)

Figure P4-7

   e. 5000-W electric heater at 240 V.

   f. Special branch circuit using three-wire No. 12 AWG in conduit to the power panel.

5. Storage

   a. Overhead lighting per Appendix Table 6 recommendation.

   b. Standard duplex outlet receptacles (12-ft spacing).

6. Nursing clinic

   a. Overhead lighting at 80-fc level.

   b. Specially designed duplex outlet receptacles for use in anesthesia (hazardous) location (12-ft spacing). Receptacles designed to accept the attachment plugs listed for use in class I, group C, NEMA type 7 hazardous locations must be used. Each branch circuit to the location must be isolated and supply power solely to that location. The enclosures must be explosion-proof, with the conductors run in threaded metal rigid conduit.

7. Elevator machinery room (E.M.)

   a. One overhead fixture and one standard receptacle.

   b. 3-hp electric motor.

8. Hallway, corridors, and linen room

   a. Overhead lighting per Appendix Table 4 recommendation.

   b. Lighting power panel location in corridor near entrance.

   c. Standard duplex outlet receptacles located as convenient.

9. Nurse, male, and female locker rest rooms (NRR, MRR, FRR)

   a. Overhead lighting at 50-fc level.

   b. Standard duplex outlet receptacles (12-ft spacing).

Note: Health care facilities require a system for emergency power and an alarm system. These will not be provided on this layout unless assigned by the instructor.

   Required: Include the following information on the layout:

   a. Complete power and lighting layout.

   b. Legend, including component symbols and conductor sizing.

   c. Compute total power requirements for lighting, receptacle, and special-purpose loads.

   d. Specify minimum service requirements based on NEC demand specifications (Appendix Table 14) and 240-V supply.

5

Summary

5-1 *NATIONAL ELECTRICAL CODE®* **SUMMARY**

All project work in this text may be completed using the theory presented herein. It is also beneficial, however, to consult the appropriate NEC Articles. This is especially necessary for special conditions and exceptions for applications other than the type covered in this text. Thus a summary of code requirement Sections will be presented.

Figure 5-1 illustrates the NEC Sections for general compliance. This includes residential, low-voltage commercial, and low-voltage industrial power applications. Other Sections may have to be used to govern particular situations. For example, Article 210 covers branch circuitry in general. A special application such as the design, layout, and installation of a fixed electric space heater requires that Section 424-3 also be consulted. Refer to Figure 5-2 for a complete listing of the specific applications and extended coverage. It will be beneficial to refer to this table when an unusual branch-circuit design problem is encountered. Other applications also require extended coverage. Outdoor branch circuits, for example, are covered in Section 225. Additional Sections are listed in Figure 5-3. If an outdoor installation involves a swimming pool, it would be considered a special case. Refer to Figure 5-3 to find the appropriate Article. In this case, it is Article 680.

With the use of electricity, safety is paramount. As mentioned previously, this is a study in itself. The NEC includes many Articles for the governance of safety. Article 240 is used for overcurrent protection. Figure 5-4 includes a list of all additional Articles used to protect equipment. Also, Article 250 is used for grounding requirements; Figure 5-5 lists special applications to other Articles.

5-2 POWER CONSUMPTION SUMMARY

Electrical power consumption can be costly. This is especially true if electricity is used for heating, since the demand is so high. The rates for heating with energy sources such as natural gas, oil, coal, and solar is generally considerably less. It is, however, quite economical to use electricity for lighting, especially fluorescent and other high-efficiency types. Also, small motor operation and appliances using electricity are clean and convenient.

Power consumption is based on the kilowatthour (kWh), as discussed in Chapter 1. Electrical energy cost is based on the amount of

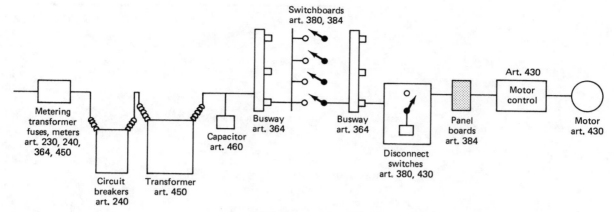

Low Voltage Industrial and Commercial Power

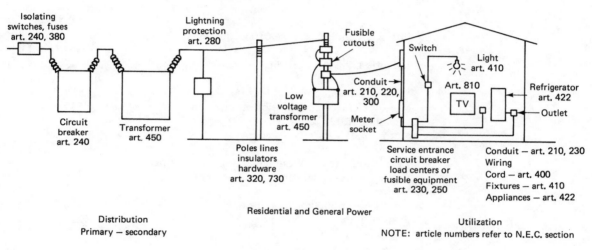

Residential and General Power

Distribution
Primary — secondary

Utilization
NOTE: article numbers refer to N.E.C. section

Figure 5-1 NEC requirements

consumption. Typically, the unit cost will decrease as more power is consumed. An illustration of this is shown in Figure 5-6. The assignments in this chapter are based on an average value since it may change several times during a month, as seen in the figure. If an average value of 10 cents/kWh is used, the cost may be figured if the power consumption level is known.

Example 1: A classroom is illuminated by forty 40-W fluorescent fixtures. Determine the average electrical power consumption cost for a 1-hour lecture.

A 40-W bulb actually consumes approximately 50 W because of the ballasts and so on. Thus the consumption for 1 hour is

$$\frac{\text{no. bulbs} \times \text{watts/bulb} \times \text{hours}}{1000 \text{ W/kWh}} = \frac{40 \text{ bulbs} \times 50 \text{ W} \times 1 \text{ hour}}{1000}$$

$$= 2.00 \text{ kWh}$$

At 10 cents/kWh, the lighting cost is 20 cents.

	NEC Article	NEC Section
Air-conditioning and refrigerating equipment		440-5
		440-31
		440-32
Busways		364-9
Class 1, class 2, and class 3 remote control, signaling, and power-limited circuits	725	
Cranes and hoists		610-42
Data processing systems		645-2
Electrical floor assemblies	366	
Electric signs and outline lighting		600-6
Electric welders	630	
Elevators, dumbwaiters, escalators, and moving walks		620-61
Fire protective signaling systems	760	
Fixed electric space heating equipment		424-3
Fixed outdoor electric deicing and snow-melting equipment		426-4
Infrared lamp industrial heating equipment		422-15
		424-3
Induction and dielectric heating equipment	665	
Marinas and boatyards		555-4
Mobile homes and mobile home parks	550	
Motion picture and television studios and similar locations	530	
Motors and motor controllers	430	
Organs		650-6
Recreational vehicles and recreational vehicle parks	551	
Sound recording and similar equipment		640-6
Switchboards and panel boards		384-22
Systems over 600 volts, nominal	710	
Systems under 50 volts	720	
Theaters and similar locations		520-41
		520-52
		520-62
X-ray equipment		660-2
		517-143

Figure 5-2 Specific-purpose branch circuits

Example 2: A residence uses 520 kWh of electricity in a 1-month period. Determine the average electrical energy cost.
The cost is

$$520 \text{ kWh} \times 10 \text{ cents/kWh} = \$52.00$$

Example 3: A commercial store uses an average of 20 A (at 240 V) per hour, 16 hours per day. Determine the cost for electricity over a 30-day period.

$$\text{kWh/day} = \frac{\text{amperes/hour} \times \text{hours} \times \text{volts}}{1000 \text{ W/kWh}}$$

$$= \frac{20 \text{ A} \times 16 \text{ hours} \times 240 \text{ V}}{1000} = 76.8 \text{ kWh}$$

	NEC *Article*
Branch circuits	210
Class 1, class 2, and class 3 remote control, signaling, and power-limited circuits	725
Communication circuits	800
Community antenna television and radio distribution systems	820
Conductors	310
Electrically driven or controlled irrigation machines	675
Electric signs and outline lighting	600
Feeders	215
Fire protective signaling systems	760
Fixed outdoor electric deicing and snow-melting equipment	426
Fixtures	410
Grounding	250
Hazardous (classified) locations	500
Hazardous (classified) locations, specific	510
Marinas and boatyards	555
Messenger supported wiring	321
Over 600 V, general	710
Overcurrent protection	240
Radio and television equipment	810
Services	230
Swimming pools, fountains, and similar installations	680
Use and identification of grounded conductors	200

Figure 5-3 Outside branch circuits and feeders

$$\text{monthly consumption} = 76.8 \text{ kWh} \times 30 \text{ days/month}$$

$$= 2364 \text{ kWh}$$

$$\text{cost} = 2364 \text{ kWh} \times 10 \text{ cents/kWh} = \$236.40$$

ASSIGNMENTS

Use an average cost of 10 cents/kWh to solve the following problems.

5-1 Determine the hourly electrical power consumption cost to light the computer-assisted design drafting lab in Assignment 2-2.

5-2 Determine the monthly electrical power consumption cost for the residential layout in Assignment 2-3 if 380 kWh is used.

5-3 Determine the monthly electrical power consumption cost for the residential layout in Assignment 2-4 if 560 kWh is used.

5-4 Determine the hourly electrical power consumption cost to light the commercial parking lot in Assignment 2-5.

5-5 Determine the monthly (30 days) electrical power consumption cost for the mall clothing store in Assignment 2-6. The store is open 12 hours per day, 7 days per week, and uses an average of 12 A (at 240 V) hourly.

	NEC Article
Air-conditioning and refrigerating equipment	440
Appliances	422
Branch circuits	210
Capacitors	460
Class 1, class 2, and class 3 remote control, signaling, and power-limited circuits	725
Cranes and hoists	610
Electrical floor assemblies	366
Electric signs and outline lighting	600
Electric welders	630
Electrolytic cells	668
Elevators, dumbwaiters, escalators, and moving walks	620
Emergency systems	700
Fire protective signaling systems	760
Fixed electric heating equipment for pipelines and vessels	427
Fixed electric space heating equipment	424
Fixed outdoor electric deicing and snow-melting equipment	426
Generators	445
Induction and dielectric heating equipment	665
Metalworking machine tools	670
Motion picture and television studios, and similar locations	530
Motors, motor circuits, and controllers	430
Organs	650
Over 600 V, nominal — general	710
Places of assembly	518
Services	230
Sound-recording and similar equipment	640
Switchboards and panelboards	384
Theaters and similar locations	520
Transformers and transformer vaults	450
X-ray equipment	517
	660

Figure 5-4   Overcurrent protection of equipment

5-6   Determine the weekly and monthly electrical cost to operate the two stores of the mall in Assignment 2-7. The beauty salon uses an average of 25 A 6 hours per day, 5 days per week. At other times the consumption rate averages 3 A per hour. The pizzeria uses an average of 20 A 12 hours per day; the consumption is negligible the remainder of the time. If there are approximately 4.35 weeks in a month, determine the average monthly cost.

5-7   Determine the average daily lighting cost for the design and drafting building in Assignment 2-8. All lights are on 10 hours per day.

	NEC Article	NEC Section
Appliances		422-16
Branch circuits		210-5
		210-6
		210-7
Cablebus		365-9
Circuits and equipment operating at less than 50 V	720	
Class 1, class 2, and class 3 circuits		725-20
		725-42
Communications circuits	800	
Community antenna television and radio distribution systems		820-7
		820-22
		820-23
Conductors	310	
Conductors (grounded)	200	
Cranes and hoists	610	
Data processing systems		645-4
Electrically driven or controlled irrigation machines		675-11(c)
		675-12
		675-13
		675-14
		675-15
Electrical floor assemblies		366-14
Electric signs and outline lighting	600	
Electrolytic cells	668	
Elevators, dumbwaiters, escalators, and moving walks	620	
Fire protective signaling systems		760-6
Fixed electric heating equipment for pipelines and vessels		427-21
		427-29
		427-48
Fixed electric space heating equipment		424-14
Fixed outdoor electric deicing and snow-melting equipment		426-27
Fixtures and lighting equipment		410-17
		410-18
		410-19
		410-21
		410-105(b)
Flexible cords		400-22
		400-23
Generators		445-8
Grounding-type receptacles (outlets)		210-7
Hazardous (classified) locations	500-517	
Health care facilities	517	
Induction and dielectric heating equipment	665	
Lighting fixtures, lampholders, lamp receptacles, and rosettes	410	
Marinas and boatyards		555-7
Metalworking machine tools and plastics	670	
Mobile homes and mobile home parks	550	
Motion picture and television studios and similar locations		530-20
		530-66
Motors, motor circuits, and controllers	430	
Organs	650	
Outlet, device, pull on and junction boxes, and fittings		370-4
		370-15

Panel boards		384-27
Radio and television equipment	810	
Receptacles and attachment plugs		410-58
Recreational vehicles and recreational vehicle parks	551	
Services	230	
Service equipment		230-63
Solar photovoltaic systems		690-41
		690-42
		690-43
		690-44
Sound recording and similar equipment		640-4
Swimming pools, fountains, and similar installations	680	
Switchboards and panel boards		384-3(c)
		384-11
Switches		380-12
Theaters and similar locations		520-81
Transformers and transformer vaults		450-10
X-ray equipment	660	517-151

Figure 5-5   Specific-purpose grounding

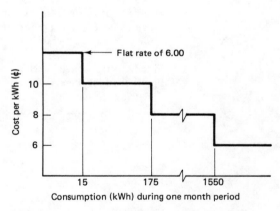

Figure 5-6   Residential power consumption cost

Appendix

Table 1 Electrical Power Consumption of Common Domestic Appliances

	approximate average wattage	approximate average kWh	circuit wiring*
Appliance			
Clock	2	17	
Small appliances			
Carving knife	100	10	
Television			
Black and white	60/160	360	
Color	100/200	400	
Refrigerator			
12 ft^3	240	730	
12 ft^3 frostless	320	1,220	
24 ft^3 frostless	610	1,830	
Freezer			
15 ft^3	340	1,200	
15 ft^3 frostless	440	1,760	
Dishwasher	1,200	360	
Special circuit			
Air conditioner			
0.75 hp	1,200	1,300	2 No. 12 (20 A)
1.5 hp	2,400	2,600	3 No. 12 (20 A)
Water heater (large)	4,500	8,000	3 No. 10 (30 A)
Dryer	4,900	1,000	3 No. 10 (30 A)
Range	12,000	1,150	3 No. 6 (50 A)†

*Plus equipment grounding wire.
†10-kW circuit.

Table 2a Common Electrical Symbols

Description	Symbol	Description	Symbol
General Outlets		Switches	
Outlet or incandescent lighting	◯	Single-pole switch	S
		Double-pole switch	S_2
		Three-way switch	S_3
Clock outlet	Ⓒ	Four-way switch	S_4
		Automatic door switch	S_D
Drop cord	Ⓓ	Electrolier switch	S_E
		Key-operated switch	S_K
Electrical outlet (use when plain circle may be confused with column or other symbols)	Ⓔ	Circuit breaker	S_{CB}
		Weatherproof circuit breaker	S_{WCB}
Fan outlet	Ⓕ	Weatherproof switch	S_{WP}
Junction box	Ⓙ	Special Outlets	
Lampholder	Ⓛ	Any standard symbol, with a lowercase subscript added, may be used for special indications. When so used, a legend on each drawing and description in specifications are strongly recommended	⬤ a, b, c etc. ⊖ a, b, c etc. $S_{a, b, c\ etc.}$
Pull switch	Ⓢ		
Outlet for vapor discharge lamp	Ⓥ	Panels, Circuits, Misc.	
		Lighting panel	▬
		Power panel	▨
		Branch circuit*: ceiling or wall	———
Exit light outlet	Ⓧ	Branch circuit*: floor	- - - -
Convenience Outlets		*Without other designation indicates 2-wire circuit.	
Duplex convenience outlet	⊖	For 3 wires	—//—
Convenience outlet other than duplex (1 = single, 3 = triplex, etc.)	⊖ 1, 3	For 4 wires	—//⁄—
Weatherproof convenience outlet	⊖ WP	Feeders (use heavy lines and designate by number for quick reference)	▬▬
Range outlet	⊖ R	Fluorescent and Exterior	
Switch and convenience outlet	⊖–S	Fluorescent fixture	◯▭ or
Special-purpose outlet (desc. in spec.)	⬤	Recessed fixture	Ⓡ▭
Floor outlet	⊙ or ▢	Exterior fixture	⬭

Table 2b Common Architectural Symbols

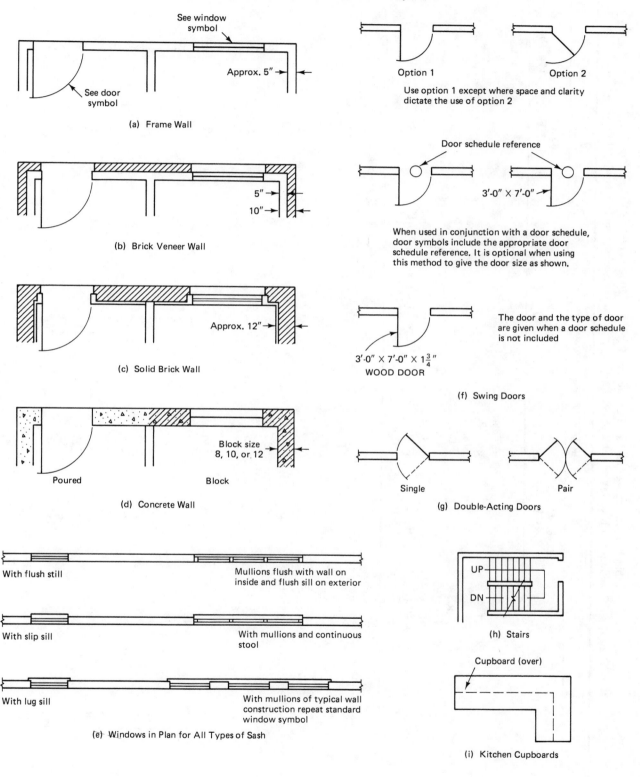

(a) Frame Wall

See window symbol

See door symbol

Approx. 5"

(b) Brick Veneer Wall

5"
10"

(c) Solid Brick Wall

Approx. 12"

(d) Concrete Wall

Poured

Block

Block size 8, 10, or 12

With flush still

With slip sill

With lug sill

(e) Windows in Plan for All Types of Sash

Mullions flush with wall on inside and flush sill on exterior

With mullions and continuous stool

With mullions of typical wall construction repeat standard window symbol

Option 1

Option 2

Use option 1 except where space and clarity dictate the use of option 2

Door schedule reference

3'-0" × 7'-0"

When used in conjunction with a door schedule, door symbols include the appropriate door schedule reference. It is optional when using this method to give the door size as shown.

3'-0" × 7'-0" × 1¾"
WOOD DOOR

The door and the type of door are given when a door schedule is not included

(f) Swing Doors

Single

Pair

(g) Double-Acting Doors

UP
DN

(h) Stairs

Cupboard (over)

(i) Kitchen Cupboards

Table 3 Ampacities of Insulated Conductors Rated 0 to 2000 V, 60 to 90°C

(A) Not More Than *Three* Conductors in Raceway or Cable or Earth (Directly Buried), Based on Ambient Temperature of 30°C (86°F)

size AWG MCM	temperature rating of conductor — copper				temperature rating of conductor — aluminum or copper-clad aluminum				size AWG MCM
	60°C (140°F) types †RUW, †T, †TW, †UF	75°C (167°F) types †FEPW, †RH, †RHW, †RUH, †THW, †THWN, †XHHW, †USE, †ZW	85°C (185°F) types V, MI	90°C (194°F) types TA, TBS, SA, AVB, SIS, †FEP, †FEPB, †RHH, †THHN, †XHHW*	60°C (140°F) types †RUW₂ †T, †TW, †UF	75°C (167°F) types †RH, †RHW, †RUH, †THW, †THWN, †XHHW, †USE	85°C (185°F) types V, MI	90°C (194°F) types TA, TBS, SA, AVB, SIS, †RHH, †THHN, †XHHW*	
18				14					18
16			18	18					16
14	20†	20†	25	25†					14
12	25†	25†	30	30†	20†	20†	25	25†	12
10	30	35†	40	40†	25†	30†	30	35†	10
8	40	50	55	55	30	40	40	45	8
6	55	65	70	75	40	50	55	60	6
4	70	85	95	95	55	65	75	75	4
3	85	100	110	110	65	75	85	85	3
2	95	115	125	130	75	90	100	100	2
1	110	130	145	150	85	100	110	115	1
0	125	150	165	170	100	120	130	135	0
00	145	175	190	195	115	135	145	150	00
000	165	200	215	225	130	155	170	175	000
0000	195	230	250	260	150	180	195	205	0000
250	215	255	275	290	170	205	220	230	250
300	240	285	310	320	190	230	250	255	300
350	260	310	340	350	210	250	270	280	350
400	280	335	365	380	225	270	295	305	400
500	320	380	415	430	260	310	335	350	500

Size								
600	355	420	460	475	285	340	370	385
700	385	460	500	520	310	375	405	420
750	400	475	515	535	320	385	420	435
800	410	490	535	555	330	395	430	450
900	435	520	565	585	355	425	465	480
1000	455	545	590	615	375	445	485	500
1250	495	590	640	665	405	485	525	545
1500	520	625	680	705	435	520	565	585
1750	545	650	705	735	455	545	595	615
2000	560	665	725	750	470	560	610	630

ampacity correction factors

For ambient temperatures over 30°C, multiply the ampacities shown above by the appropriate correction factor to determine the maximum allowable load current.

ambient temp. (°C)									ambient temp. (°F)
31-40	0.82	0.88	0.90	0.91	0.82	0.88	0.90	0.91	86-104
41-45	0.71	0.82	0.85	0.87	0.71	0.82	0.85	0.87	105-113
46-50	0.58	0.75	0.80	0.82	0.58	0.75	0.80	0.82	114-122
51-60		0.58	0.67	0.71		0.58	0.67	0.71	123-141
61-70		0.35	0.52	0.58		0.35	0.52	0.58	142-158
71-80			0.30	0.41			0.30	0.41	159-176

† The overcurrent protection for conductor types marked with an obelisk (†) shall not exceed 15 A for 14 AWG, 20 A for 12 AWG, and 30 A for 10 AWG copper; or 15 A for 12 AWG and 25 A for 10 AWG aluminum and copper-clad aluminum after any correction factors for ambient temperature and number of conductors have been applied.

*For dry locations only. See 75°C column for wet locations.

Table 3 *(continued)*

(B) *Single* Conductors in Free Air, Based on Ambient Temperature of 30°C (86°F)

temperature rating of conductor

copper / *aluminum or copper-clad aluminum*

AWG MCM	60°C (140°F) types †RUW, †T, †TW	75°C (167°F) types †FEPW, †RH, †RHW, †RUH, †THW, †THWN, †XHHW, †ZW	85°C (185°F) types V, MI	90°C (194°F) types TA, TBS, SA, AVB, SIS, †FEP, †FEPB, †RHH, †THHN, †XHHW*	60°C (140°F) types †RUW, †T, †TW	75°C (167°F) types †RH, †RHW, †RUH, †THW, †THWN, †XHHW	85°C (185°F) types V, MI	90°C (194°F) types TA, TBS, SA, AVB, SIS, †RHH, †THHN, †XHHW*	AWG MCM
18				18					
16			23	24					
14	25†	30†	30	35†					
12	30†	35†	40	40†	25†	30†	30	35†	12
10	40†	50†	55	55†	35†	40†	40	40†	10
8	60	70	75	80	45	55	60	60	8
6	80	95	100	105	60	75	80	80	6
4	105	125	135	140	80	100	105	110	4
3	120	145	160	165	95	115	125	130	3
2	140	170	185	190	110	135	145	150	2
1	165	195	215	220	130	155	165	175	1
0	195	230	250	260	150	180	195	205	0
00	225	265	290	300	175	210	225	235	00
000	260	310	335	350	200	240	265	275	000
0000	300	360	390	405	235	280	305	315	0000
250	340	405	440	455	265	315	345	355	250
300	375	445	485	505	290	350	380	395	300
350	420	505	550	570	330	395	430	445	350
400	455	545	595	615	355	425	465	480	400
500	515	620	675	700	405	485	525	545	500

600	575	690	750	780	455	540	595	615	600
700	630	755	825	855	500	595	650	675	700
750	655	785	855	885	515	620	675	700	750
800	680	815	885	920	535	645	700	725	800
900	730	870	950	985	580	700	760	785	900
1000	780	935	1020	1055	625	750	815	845	1000
1250	890	1065	1160	1200	710	855	930	960	1250
1500	980	1175	1275	1325	795	950	1035	1075	1500
1750	1070	1280	1395	1445	875	1050	1145	1185	1750
2000	1155	1385	1505	1560	960	1150	1250	1335	2000

ampacity correction factors

For ambient temperatures over 30°C, multiply the ampacities shown above by the appropriate correction factor shown below.

ambient temp. (°C)									ambient temp. (°F)
31-40	0.82	0.88	0.90	0.91	0.82	0.88	0.90	0.91	86-104
41-45	0.71	0.82	0.85	0.87	0.71	0.82	0.85	0.87	105-113
46-50	0.58	0.75	0.80	0.82	0.58	0.75	0.80	0.82	114-122
51-60		0.58	0.67	0.71		0.58	0.67	0.71	123-141
61-70		0.35	0.52	0.58		0.35	0.52	0.58	142-158
71-80			0.30	0.41			0.30	0.41	159-176

†The overcurrent protection for conductor types marked with an obelisk (†) shall not exceed 20 A for 14 AWG, 25 A for 12 AWG, and 40 A for 10 AWG copper, or 20 A for 12 AWG and 30 A for 10 AWG aluminum and copper-clad aluminum after any correction factor for ambient has been applied.
*For dry locations only. See 75°C column for wet locations.

Source: Reprinted with permission from NFPA70-1984, *National Electrical Code®*, Copyright © 1983, National Fire Protection Association, Boston, Mass. This reprinted material is not the complete and official position of the NFPA on the referenced subject, which is represented only by the standard in its entirety.

Table 4   General Lighting Loads by Occupancies

type of occupancy	unit load per square foot (W) (Volt-Amperes)
Armories and auditoriums	1
Banks	3½[†]
Barber shops and beauty parlors	3
Churches	1
Clubs	2
Courtrooms	2
Dwelling units[‡]	3
Garages – commercial (storage)	½
Hospitals	2
Hotels and motels, including apartment houses without provisions for cooking by tenants[‡]	2
Industrial commercial (loft) buildings	2
Lodge rooms	1½
Office buildings	3½[†]
Restaurants	2
Schools	3
Stores	3
Warehouses (storage)	¼
In any of the above occupancies except one-family dwellings and individual dwelling units of two-family and multifamily dwellings:	
Assembly halls and auditoriums	1
Halls, corridors, closets, stairways	½
Storage spaces	¼

*For SI units: 1 ft^2 = 0.093 m^2.

[†] In addition a unit load of 1 VA/ft^2 shall be included for general-purpose receptacle outlets when the actual number of general-purpose receptacle outlets is unknown.

[‡] All receptacle outlets of 20 A or less rating in one-family, two-family, and multifamily dwellings and in guest rooms of hotels and motels [except those connected to the receptacle circuits specified in Section 220-3(b)] shall be considered as outlets for general illumination, and no additional load calculations shall be required for such outlets.

Source: Reprinted with permission from NFPA70-1984, *National Electrical Code*®, Copyright © 1983, National Fire Protection Association, Boston, Mass. This reprinted material is not the complete and official position of the NFPA on the referenced subject, which is represented only by the standard in its entirety.

Table 5 Initial Lumen Ratings for Common Interior Lamps

	nominal wattage	*initial lumens*
Fluorescent	30 (36 in. length)	2,400
	40 (48 in. length)	3,100
	80 (96 in. length)	6,300
Mercury vapor	75	2,650
	100	3,800
	175	7,300
	250	11,000
Incandescent	60	855
	100	1,750
	150	2,730
	200	3,940
	300	6,240
	250 (floodlight)	3,750
	400 (floodlight)	6,700

Table 6 Recommended Interior Footcandle Levels

	*footcandle level**
Bakeries	50
Bathrooms and washrooms:	
General	10-20
At mirror	30-50
Control rooms	50
Corridors and stairways	10-20
Drafting	120
Entrance and lobbies	10-30
Material handling	30-50
Offices:	
Reading/transcribing	70
Regular office work	80-100
Parking lots (exterior)	1-4
Restaurants:	
Cashier	50
Intimate dining	10
Leisure environment	30-50
Kitchen	30-70
Storage rooms and warehouses	10-30
Stores:	
Feature display	500
General show window (200 W/linear ft min)	100-200
Merchandising area	100

*These are general levels and vary within the range given, depending on specific application.

Appendix

Table 7 Coefficient of Utilization*

(A) Cavity Ratio (CR) Number

room dimensions		distance from fixture to work surface (ft)				
width	length	2.0	4.0	6.0	8.0	12.0
10	10	2.0	4.0	6.0	8.0	
	20	1.5	3.0	4.5	6.0	9.0
	30	1.3	2.7	4.0	5.3	8.0
	40	1.2	2.5	3.7	5.0	7.5
	60	1.2	2.3	3.5	4.7	7.1
20	20	1.0	2.0	3.0	4.0	6.0
	30	0.8	1.7	2.5	3.3	4.9
	60	0.7	1.3	2.0	2.7	4.0
	90	0.6	1.2	1.8	2.4	3.6
30	30	0.7	1.3	2.0	2.7	4.0
	45	0.6	1.1	1.7	2.2	3.3
	60	0.5	1.0	1.5	2.0	3.0
	90	0.4	0.9	1.3	1.8	2.7
	150	0.4	0.8	1.2	1.6	2.4
42	42	0.5	1.0	1.4	1.9	2.8
	60	0.4	0.8	1.2	1.6	2.4
	90	0.3	0.7	1.0	1.4	2.1
	140	0.3	0.6	0.9	1.2	1.9
	200	0.3	0.6	0.9	1.1	1.7

(B) Coefficient of Utilization

CR	80% ceiling reflectance		50% ceiling reflectance	
	50% walls	30% walls	50% walls	30% walls
1	0.70	0.66	0.62	0.59
2	0.60	0.54	0.53	0.49
3	0.50	0.46	0.46	0.41
4	0.46	0.39	0.41	0.36
5	0.40	0.33	0.36	0.30
6	0.36	0.29	0.32	0.26
7	0.32	0.25	0.29	0.23
8	0.29	0.22	0.26	0.20
9	0.26	0.19	0.23	0.18
10	0.23	0.17	0.21	0.16

*Based on the use of a common-type luminaire.

Table 8a Illumination Ratings for Common H.I.D. Lamps

type	nominal wattage	approximate initial lumens	footcandle level	approximate spacing (ft)
Mercury Vapor	175	7,300	1.0*	20×20
			0.4*	60×60
	250	11,000	1.0*	20×20
			0.6*	60×60
			0.3*	80×80
	400	20,000	1.0†	140×140
			1.5†	120×120
			2.0†	100×100
	1,000	56,000	1.5†	175×175
			2.0†	150×150
			3.0†	125×125
			4.0†	120×100
			1.5‡	120×140
			2.0‡	120×100
Multivapor	400	31,500 ⎤		
	1,000	90,000	Refer to Table 8(b) for fixture spacing using the watts per square foot method.	
Sodium	250	30,000		
	400	50,000		
	1,000	140,000 ⎦		

*One lamp per pole (20-ft mounting).
† One lamp per pole (30-ft mounting).
‡ Two lamps per pole (30-ft mounting).

Table 8b Illumination Ratings for Common H.I.D. Lamps (continued)

Location	Footcandle Level	Watts per sq. ft.			
		Mercury Vapor	Metal Halide	High Pressure Sodium	Low Pressure Sodium
Building					
General construction	10	0.6			
Excavation work	2	.14			
Floodlight, building exteriors and monuments	15	2.5			
Bulletin and poster board surfaces	50	7.0			
Central station					
Catwalks	2	.4			
Coal unloading dock	5	.8			
Coal storage area	0.1	.01			
Conveyors	2	.25			
Entrances, general or service building					
Main	10	1.0			
Secondary	2	.26			
Oil storage tanks	1	.1			
Open yard	0.2	.02			
Platforms − boiler, turbine deck	5	1	Multiply Mercury Watts by .7	Multiply Mercury Watts by .4	Multiply Mercury Watts by .3
Roadway					
Between or along buildings	1	.15			
Not bordered by buildings	0.5	.05			
Gardens					
General lighting	0.5	.05			
Path, steps, away from house	1	.15			
Lumber yards	1	.1			
Parking lots					
Large − over 15,000 sq.ft.	5	.3			
	2	.15			
	1	.08			
	5	.5			
Small − up to 15,000 sq.ft.	2	.23			
	1	.2			
Protective lighting − Entrances					
Active (pedestrian/conveyance)	5	.75			
Inactive (infrequently used)	1	.13			
Vital locations or structures	5	1.0			
Building surrounds	1	.12			
Storage areas − active	20	3			
Storage areas − inactive	1	.1			
Loading and unloading platforms	20	3			
General inactive areas	0.20	.028			
Service station (at grade)					
Approach	1.5	.20			
Driveway	1.5	.20			
Pump island area	20	3			
Building faces (exclusive of glass)	10	1.0			
Service areas	3	.5			

Table 9a Single-Phase Motor Ratings*

hp	*115 V*	*230 V*
1/6	4.4	2.2
1/4	5.8	2.9
1/3	7.2	3.6
1/2	9.8	4.9
3/4	13.8	6.9
1	16	8
1 1/2	20	10
2	24	12
3	34	17
5	56	28
7 1/2	80	40
10	100	50

*The following values of full-load currents are for motors running at usual speeds and motors with normal torque characteristics. Motors built for especially low speeds or high torques may have higher full-load currents, and multispeed motors will have full-load current varying with speed, in which case the nameplate current ratings shall be used.

To obtain full-load currents of 208- and 200-V motors, increase corresponding 230-V motor full-load currents by 10 and 15%, respectively.

The voltages listed are rated motor voltages. The currents listed shall be permitted for system voltage ranges of 110 to 120 and 220 to 240.

Source: Reprinted with permission from NFPA70-1984, *National Electrical Code*®, Copyright © 1983, National Fire Protection Association, Boston, Mass. This reprinted material is not the complete and official position of the NFPA on the referenced subject, which is represented only by the standard in its entirety.

Table 9b Three-Phase Motor Ratings*

| | full-load current[†] three-phase alternating-current motors | | | | | | | | | transformer kVA rating | |
| | induction-type squirrel cage and wound-rotor (A) | | | | | synchronous-type unity power factor[‡] (A) | | | | induction-type squirrel-cage and wound-rotor[§] (kVA) | |
hp	115 V	230 V	460 V	575 V	2300 V	230 V	460 V	575 V	2300 V	230 V	460 V
0.50	4	2	1	.8						2	1
0.75	5.6	2.8	1.4	1.1						2.8	1.4
1	7.2	3.6	1.8	1.4						3.5	1.8
1.50	10.4	5.2	2.6	2.1						5	2.5
2	13.6	6.8	3.4	2.7						6.5	3.3
3		9.6	4.8	3.9						9	4.5
5		15.2	7.6	6.1						15	7.5
7.50		22	11	9						22	11
10		28	14	11						27	14
15		42	21	17						40	20
20		54	27	22						52	26
25		68	34	27		53	26	21		64	32
30		80	40	32		63	32	26		78	39
40		104	52	41		83	41	33		104	52
50		130	65	52		104	52	42		125	63
60		154	77	62	16	123	61	49	12		
75		192	96	77	20	155	78	62	15		
100		248	124	99	26	202	101	81	20		
125		312	156	125	31	253	126	101	25		
150		360	180	144	37	302	151	121	30		
200		480	240	192	49	400	201	161	40		

*For full-load currents of 208- and 200-V motors, increase the corresponding 230-V motor full-load by 10 and 15% respectively.

[†] These values of full-load current are for motors running at speeds usual for belted motors and motors with normal torque characteristics. Motors built for especially low speeds or high torques may require more running current, and multispeed motors will have full-load current varying with speed, in which case the nameplate current rating shall be used.

[‡] For 90% and 80% power factor, the figures should be multiplied by 1.1 and 1.25, respectively.

The voltages listed are rated motor voltages. The currents listed shall be permitted for system voltage ranges of 110 to 120, 220, 440 to 480, and 550 to 600 V.

[§] For estimating purposes only, allow 20% additional kVA if motors are started more than once per hour.

Source: Reprinted with permission from NFPA70-1984, *National Electrical Code*®, Copyright © 1983, National Fire Protection Association, Boston, Mass. This reprinted material is not the complete and official position of the NFPA on the referenced subject, which is represented only by the standard in its intirety.

Table 10   Three-Phase Motor Circuit Sizes*

hp	full-load current	switch/ fuse	circuit breaker trip	starter size	conductor size†	conduit diameter (in.)
1	1.8	30/2	15	1	3 No. 12	0.75
1.5	2.6	30/2.8	15	1	3 No. 12	0.75
2	3.4	30/4	15	1	3 No. 12	0.75
3	4.8	30/5.6	15	1	3 No. 12	0.75
5	7.6	30/9	20	1	3 No. 12	0.75
7.5	11	30/15	30	1	3 No. 12	0.75
10	14	30/17.5	30	1	3 No. 12	0.75
15	21	30/25	50	2	3 No. 10	0.75
20	27	60/35	60	2	3 No. 8	0.75
25	34	60/40	70	2	3 No. 6	1
30	40	60/45	90	3	3 No. 6	1
40	52	100/60	125	3	3 No. 4	1.25
50	65	100/80	150	3	3 No. 2	1.25
60	77	200/90	175	4	3 No. 1	1.50
75	96	200/110	200	4	3 No. 1	1.50
100	124	200/150	250	4	3 No. 2/0	2
125	156	400/200	300	5	3 No. 3/0	2
150	180	400/225	350	5	3 No. 4/0	2.5
200	240	400/300	400	5	3 No. 350 MCM	3

*This table has been calculated based on 460 V, three-phase, 60-Hz motors using type XHHW copper conductor and 75° Celsius temperature rating. Since the branch circuits are commonly long runs, the sizes in this table generally exceed minimum requirements. They are commonly used by industrial electrical designers.
† For wire runs up to 200 ft.

Table 11 Recommended Conductor Size for Long Distribution Runs*

circuit length (ft) for 3% voltage drop at:

size AWG/MCM	15 A	20 A	25 A	35 A	50 A	70 A	80 A	90 A	100 A	125 A	150 A	175 A	250 A	300 A
14	42													
12	66	50												
10	105	79	63											
8	168	126	100	72										
6	267	200	160	114	90									
4	424	318	255	182	127	91								
3	536	402	321	229	160	114	100							
2	679	507	405	289	202	144	126	112						
1	852	639	511	365	255	182	159	142	127					
0	1,074	806	644	460	322	230	201	179	161	128				
00	1,355	1,016	813	581	405	290	254	225	203	162	135			
000	1,709	1,281	1,025	732	512	366	320	254	256	205	170	146		
0,000	2,155	1,616	1,293	923	646	461	404	359	323	258	215	184		
250,000	2,546	1,911	1,527	1,091	763	545	477	424	381	305	254	218	152	
300,000	3,055	2,291	1,833	1,309	916	654	572	509	458	366	305	261	183	
350,000		2,673	2,138	1,526	1,069	763	667	594	534	427	356	305	213	178
400,000		3,055	2,444	1,746	1,222	873	763	679	611	488	407	349	244	203
500,000			3,055	2,182	1,527	1,091	934	848	763	611	509	436	305	254
600,000				2,619	1,833	1,309	1,145	1,018	916	733	611	523	366	305
700,000				3,055	2,138	1,527	1,336	1,188	1,069	855	712	611	427	356

*Table is calculated for 120 V. The lengths (in feet) shown are approximate for single-phase and two-phase at unity power factor. For three-phase the footage may be increased by approximately 12%.

Table 12 Copper Conductor Information

AWG no.	typical number of strands	nominal diameter (in.)	approximate breaking strength (hard drawn) (lb)	approximate resistance (Ω/1000 ft)
14	1	0.064	190	2.5
12	1	0.081	300	1.6
10	1	0.102	490	1.0
8	1	0.128	780	0.64
6	7	0.184	1230	0.41
4	7	0.232	1940	0.26
3	7	0.260	2440	0.20
2	7	0.292	3050	0.16
1	7 or 19	0.332		0.13
0	7 or 19	0.372		0.10
00	7 or 19	0.418		0.08
000	19	0.470		0.06
0000	19	0.528		0.05

Table 13 Sample Title Block and Bill of Material

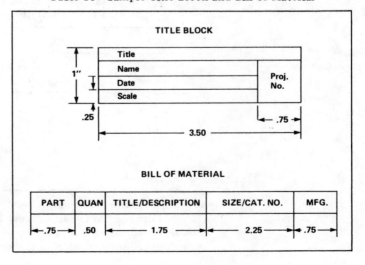

Table 14   Lighting-Load Feeder Demand Factors

type of occupancy	portion of lighting load to which demand factor applies (VA)	demand factor (%)
Dwelling units	First 3000 or less at	100
	From 3001 to 120,000 at	35
	Remainder over 120,000 at	25
Hospitals*	First 50,000 or less at	40
	Remainder over 50,000 at	20
Hotels and motels,* including apartment houses without provision for cooking by tenants	First 20,000 or less at	50
	From 20,001 to 100,000 at	40
	Remainder over 100,000 at	30
Warehouses (storage)	First 12,500 or less at	100
	Remainder over 12,500 at	50
All others	Total Volt-amperes	100

*The demand factors of this table shall not apply to the computed load of feeders to areas in hospitals, hotels, and motels where the entire lighting is likely to be used at one time, as in operating rooms, ballrooms, or dining rooms.

Source: Reprinted with permission from NFPA70-1984, *National Electrical Code*®, Copyright © 1983, National Fire Protection Association, Boston, Mass. This reprinted material is not the complete and official position of the NFPA on the referenced subject, which is represented only by the standard in its entirety.

Table 15   Conductor Types and Sizes, RH-RHH-RHW-THW-THWN-THHN-XHHW

AWG		service rating (A)
copper	aluminum and copper-clad aluminum	
4	2	100
3	1	110
2	1/0	125
1	2/0	150
1/0	3/0	175
2/0	4/0	200

Source: Reprinted with permission from NFPA70-1984, *National Electrical Code*®, Copyright © 1983, National Fire Protection Association, Boston, Mass. This reprinted material is not the complete and official position of the NFPA on the referenced subject, which is represented only by the standard in its entirety.

Table 16 Maximum Number of Conductors in Conduit or Tubing

insulation type letters	conductor size AWG/MCM	conduit trade size (in.)												
		½	¾	1	1¼	1½	2	2½	3	3½	4	4½	5	6
TW, T, RUH, RUW, XHHW (14 through 8)	14	9	15	25	44	60	99	142	171					
	12	7	12	19	35	47	78	111	131	176				
	10	5	9	15	26	36	60	85						
	8	2	4	7	12	17	28	40	62	84	108			
RHW and RHH (without outer covering), THW	14	6	10	16	29	40	65	93	143	192				
	12	4	8	13	24	32	53	76	117	157	163			
	10	4	6	11	19	26	43	61	95	127				
	8	1	3	5	10	13	22	32	49	66	85	106	133	
TW, T, THW, RUH (6 through 2), RUW (6 through 2)	6	1	2	4	7	10	16	23	36	48	62	78	97	141
	4	1	1	3	5	7	12	17	27	36	47	58	73	106
	3	1	1	2	4	6	10	15	23	31	40	50	63	91
	2	1	1	2	4	5	9	13	20	27	34	43	54	78
	1		1	1	3	4	6	9	14	19	25	31	39	57
FEPB (6 through 2), RHW and RHH (without outer covering)	0		1	1	2	3	5	8	12	16	21	27	33	49
	00		1	1	1	3	5	7	10	14	18	23	29	41
	000		1	1	1	2	4	6	9	12	15	19	24	35
	0000			1	1	1	3	5	7	10	13	16	20	29
	250			1	1	1	2	4	6	8	10	13	16	23
	300			1	1	1	2	3	5	7	9	11	14	20
	350				1	1	1	3	4	6	8	10	12	18
	400				1	1	1	2	4	5	7	9	11	16
	500				1	1	1	1	3	4	6	7	9	14
	600					1	1	1	3	4	5	6	7	11
	700					1	1	1	2	3	4	5	7	10
	750					1	1	1	2	3	4	5	6	9
THWN	14	13	24	39	69	94	154							
	12	10	18	29	51	70	114	164						
	10	6	11	18	32	44	73	104	160					
	8	3	5	9	16	22	36	51	79	106	136			

Table 16 *(continued)*

insulation type letters	conductor size AWG/MCM	conduit trade size (in.)												
		½	¾	1	1¼	1½	2	2½	3	3½	4	4½	5	6
THHN, FEP (14 through 2), FEPB (14 through 8), PFA (14 through 4/0), PFAH (14 through 4/0), Z (14 through 4/0)	6	1	4	6	11	15	26	37	57	76	98	125	154	
	4	1	2	4	7	9	16	22	35	47	60	75	94	137
	3	1	1	3	6	8	13	19	29	39	51	64	80	116
	2	1	1	3	5	7	11	16	25	33	43	54	67	97
	1		1	1	3	5	8	12	18	25	32	40	50	72
XHHW (4 through 500MCM)	0		1	1	3	4	7	10	15	21	27	33	42	61
	00		1	1	2	3	6	8	13	17	22	28	35	51
	000		1	1	1	3	5	7	11	14	18	23	29	42
	0000		1	1	1	2	4	6	9	12	15	19	24	35
	250			1	1	1	3	4	7	10	12	16	20	28
	300			1	1	1	3	4	6	8	11	13	17	24
	350			1	1	1	2	3	5	7	9	12	15	21
	400				1	1	1	3	5	6	8	10	13	19
	500				1	1	1	2	4	5	7	9	11	16
	600				1	1	1	1	3	4	5	7	9	13
	700					1	1	1	3	4	5	6	8	11
	750					1	1	1	2	3	4	6	7	11
RHW	14	3	6	10	18	25	41	58	90	121	155			
	12	3	5	9	15	21	35	50	77	103	132			
	10	2	4	7	13	18	29	41	64	86	110	138		
	8	1	2	4	7	9	16	22	35	47	60	75	94	137
RHH (with outer covering)	6	1	1	2	5	6	11	15	24	32	41	51	64	93
	4	1	1	1	3	5	8	12	18	24	31	39	50	72
	3	1	1	1	3	4	7	10	16	22	28	35	44	63
	2		1	1	3	4	6	9	14	19	24	31	38	56
	1		1	1	1	3	5	7	11	14	18	23	29	42
	0		1	1	1	2	4	6	9	12	16	20	25	37
	00			1	1	1	3	5	8	11	14	18	22	32
	000			1	1	1	3	4	7	9	12	15	19	28
	0000			1	1	1	2	4	6	8	10	13	16	24

250	1	1	3	5	6	8	11	13	19
300	1	1	3	4	5	7	9	11	17
350	1	1	2	4	5	6	8	10	15
400	1	1	1	3	4	6	7	9	14
500	1	1	1	3	4	5	6	8	11
600	1	1	1	2	3	4	5	6	9
700	1	1	1	1	3	3	4	6	8
750		1	1	1	3	3	4	5	8

Source: Reprinted with permission from NFPA70-1984, *National Electrical Code*®, Copyright © 1983, National Fire Protection Association, Boston, Mass. This reprinted material is not the complete and official position of the NFPA on the referenced subject, which is represented only by the standard in its entirety.

Appendix

Table 17 Conduit Bend Information*

size of conduit† (in.)	NEC for conductors without lead sheath
0.50	4
0.75	5
1.00	6
1.25	8
1.50	10
2.00	12
2.50	15
3.00	18
3.50	21
4.00	24
5.00	30
6.00	36

*A run of conduit shall not contain more than 360° of total bend.
† Minimum inside radius of conduit.

Table 18 Dimensions and Percent Area of Conduit and Tubing*

			area (in.²)							
			not lead covered			lead covered				
trade size	internal diameter (in.)	total 100%	2 cond. 31%	over 2 cond. 40%	1 cond. 53%	1 cond. 55%	2 cond. 30%	3 cond. 40%	4 cond. 38%	over 4 cond. 35%
½	0.622	0.30	0.09	0.12	0.16	0.17	0.09	0.12	0.11	0.11
¾	0.824	0.53	0.16	0.21	0.28	0.29	0.16	0.21	0.20	0.19
1	1.049	0.86	0.27	0.34	0.46	0.47	0.26	0.34	0.33	0.30
1¼	1.380	1.50	0.47	0.60	0.80	0.83	0.45	0.60	0.57	0.53
1½	1.610	2.04	0.63	0.82	1.08	1.12	0.61	0.82	0.78	0.71
2	2.067	3.36	1.04	1.34	1.78	1.85	1.01	1.34	1.28	1.18
2½	2.469	4.79	1.48	1.92	2.54	2.63	1.44	1.92	1.82	1.68
3	3.068	7.38	2.29	2.95	3.91	4.06	2.21	2.95	2.80	2.58
3½	3.548	9.90	3.07	3.96	5.25	5.44	2.97	3.96	3.76	3.47
4	4.026	12.72	3.94	5.09	6.74	7.00	3.82	5.09	4.83	4.45
4½	4.506	15.94	4.94	6.38	8.45	8.77	4.78	6.38	6.06	5.56
5	5.047	20.00	6.20	8.00	10.60	11.00	6.00	8.00	7.60	7.00
6	6.065	28.89	8.96	11.56	15.31	15.89	8.67	11.56	10.98	10.11

*Areas of conduit or tubing for the combinations of wires permitted in Table 1, Chapter 9, of the NEC.

Source: Reprinted with permission from NFPA70-1984, *National Electrical Code*®, Copyright © 1983, National Fire Protection Association, Boston, Mass. This reprinted material is not the complete and official position of the NFPA on the referenced subject, which is represented only by the standard in its entirety.

Table 19 Conductor Dimensions

(A) Dimensions of Rubber-Covered and Thermoplastic-Covered Conductors

size AWG MCM	types RFH-2, RH, RHH,*** RHW,*** SF-2		types TF, T, THW,† TW, RUH,** RUW**		types TFN, THHN, THWN		types**** FEP, FEPB, FEPW, TFE, PF, PFA, PFAH, PGF, PTF, Z, ZF, ZFF		type XHHW, ZW††		types KF-1, KFF-1, KF-2, KFF-2	
	approx. diam. (in.)	approx. area (in.²)	approx. diam. (in.)	approx. area (in.²)	approx. diam. (in.)	approx. area (in.²)	approx. diam. (in.)	approx. area (in.²)	approx. diam. (in.)	approx. area (in.²)	approx. diam. (in.)	approx. area (in.²)
col. 1	col. 2	col. 3	col. 4	col. 5	col. 6	col. 7	col. 8	col. 9	col. 10	col. 11	col. 12	col. 13
18	.146	.0167	.106	.0088	.089	.0062	.081	.0052			.065	.0033
16	.158	.0196	.118	.0109	.100	.0079	.092	.0066			.070	.0038
14	30 mils .171	.0230	.131	.0135	.105	.0087	.105 .105	.0087 .0087			.083	.0054
14	45 mils .204*	.0327*										
12	30 mils .188	.0278	.162† .148	.0206† .0172	.122	.0117	.121 .121	.0115 .0115	.129	.0131	.102	.0082
12	45 mils .221*	.0384*										
10	.242	.0460	.179† .168 .199†	.0252† .0222 .0311†	.153	.0184	.142 .142	.0158 .0158	.146	0.167	.124	.0121
8	.328	.0845	.245	.0471†	.218	.0373	.206	.0333	.166	.0216		
8			.276†	.0598†			.186	.0272	.241	.0456		
6	.397	.1238	.323	.0819	.257	.0519	.244 .302	.0468 .0716	.282	.0625		
4	.452	.1605	.372	.1087	.328	.0845	.292 .350	.0669 .0962	.328	.0845		
3	.481	.1817	.401	.1263	.356	.0995	.320 .378	.0804 .1122	.356	.0995		
2	.513	.2067	.433	.1473	.388	.1182	.352 .410	.0973 .1320	.388	.1182		
1	.588	.2715	.508	.2027	.450	.1590	.420	.1385	.450	.1590		
0	.629	.3107	.549	.2367	.491	.1893	.462	.1676	.491	.1893		
00	.675	.3578	.595	.2781	.537	.2265	.498	.1948	.537	.2265		
000	.727	.4151	.647	.3288	.588	.2715	.560	.2463	.588	.2715		
0000	.785	.4840	.705	.3904	.646	.3278	.618	.2999	.646	.3278		

(A) *(continued)*

size AWG MCM	types RFH-2, RH, RHH,*** RHW,*** SF-2		types TF, T, THW,† TW, RUH,** RUW**		types TFN, THHN, THWN		types**** FEP, FEPB, FEPW, TFE, PF, PFA, PFAH, PGF, PTF, Z, ZF, ZFF		type XHHW, ZW††	
	approx. diam. (in.)	approx. area (in.²)	approx. diam. (in.)	approx. area (in.²)	approx. diam. (in.)	approx. area (in.²)	approx. diam. (in.)	approx. area (in.²)	approx. diam. (in.)	approx. area (in.²)
col. 1	col. 2	col. 3	col. 4	col. 5	col. 6	col. 7	col. 8	col. 9	col. 10	col. 11
250	0.868	0.5917	0.788	0.4877	0.716	0.4026			0.716	0.4026
300	0.933	0.6837	0.843	0.5581	0.771	0.4669			0.771	0.4669
350	0.985	0.7620	0.895	0.6291	0.822	0.5307			0.822	0.5307
400	1.032	0.8365	0.942	0.6969	0.869	0.5931			0.869	0.5931
500	1.119	0.9834	1.029	0.8316	0.955	0.7163			0.955	0.7163
600	1.233	1.1940	1.143	1.0261	1.058	0.8791			1.073	0.9043
700	1.304	1.3355	1.214	1.1575	1.129	1.0011			1.145	1.0297
750	1.339	1.4082	1.249	1.2252	1.163	1.0623			1.180	1.0936
800	1.372	1.4784	1.282	1.2908	1.196	1.1234			1.210	1.1499
900	1.435	1.6173	1.345	1.4208	1.259	1.2449			1.270	1.2668
1000	1.494	1.7530	1.404	1.5482	1.317	1.3623			1.330	1.3893
1250	1.676	2.2062	1.577	1.9532					1.500	1.7671
1500	1.801	2.5475	1.702	2.2751					1.620	2.0612
1750	1.916	2.8832	1.817	2.5930					1.740	2.3779
2000	2.021	3.2079	1.922	2.9013					1.840	2.6590

*The dimensions of Types RHH and RHW.

**No. 14 to No. 2.

†Dimensions of THW in sizes No. 14 to No. 8. No. 6 THW and larger is the same dimension as T.

***Dimensions of RHH and RHW without outer covering are the same as THW No. 18 to No. 10, solid; No. 8 and larger, stranded.

****In Columns 8 and 9 are for FEPB, Z, ZF, and ZFF only.

*****In Columns 8 and 9 the values shown for sizes No. 1 thru .0000 are for TFE and Z only. The right-hand values in Columns 8 and 9 are for FEPB, Z, ZF, and ZFF only.

††No. 14 to No. 2.

(B) Dimensions of Lead-Covered Conductors, Types RL, RHL, and RUL[†††]

size AWG-MCM	single conductor		two conductors		three conductors	
	diam. (in.)	area (in.²)	diam. (in.)	area (in.²)	diam. (in.)	area (in.²)
14	0.28	0.062	0.28 × 0.47	0.115	0.59	0.273
12	0.29	0.066	0.31 × 0.54	0.146	0.62	0.301
10	0.35	0.096	0.35 × 0.59	0.180	0.68	0.363
8 sol.	0.41	0.132	0.41 × 0.71	0.255	0.82	0.528
8 str.	0.43	0.145	0.43 × 0.75	0.282	0.86	0.581
6	0.49	0.188	0.49 × 0.86	0.369	0.97	0.738
4	0.55	0.237	0.54 × 0.96	0.457	1.08	0.916
2	0.60	0.283	0.61 × 1.08	0.578	1.21	1.146
1	0.67	0.352	0.70 × 1.23	0.756	1.38	1.49
0	0.71	0.396	0.74 × 1.32	0.859	1.47	1.70
00	0.76	0.454	0.79 × 1.41	0.980	1.57	1.94
000	0.81	0.515	0.84 × 1.52	1.123	1.69	2.24
0000	0.87	0.593	0.90 × 1.64	1.302	1.85	2.68
250	0.98	0.754			2.02	3.20
300	1.04	0.85			2.15	3.62
350	1.10	0.95			2.26	4.02
400	1.14	1.02			2.40	4.52
500	1.23	1.18			2.59	5.28

[†††] These cables are limited to straight runs or with nominal offsets equivalent to not more than two quarter bends.

Note: No. 14 to No. 10, solid conductors; No. 8, solid or stranded conductors; No. 6 and larger, stranded conductors.

Source: Reprinted with permission from NFPA70-1984, *National Electrical Code®*, Copyright © 1983, National Fire Protection Association, Boston, Mass. This reprinted material is not the complete and official position of the NFPA on the referenced subject, which is represented only by the standard in its entirety.

Table 20 Allowable Cable Fill Area

(A) Multiconductor Cables in Ladder, Ventilated Trough, or Solid-Bottom
Cable Trays for Cables Rated 2000 V or Less

| | *maximum allowable fill area in square inches for multiconductor cables*[†] | | | |
| | *ladder or ventilated-trough cable trays* | | *solid-bottom cable trays* | |
inside width of cable tray (in.)	column 1 only (in.^{2})	column 2* only (in.^{2})	column 3 only (in.^{2})	column 4* only (in.^{2})
6	7	7−(1.2 Sd)**	5.5	5.5−Sd**
12	14	14−(1.2 Sd)	11.0	11.0−Sd
18	21	21−(1.2 Sd)	16.5	16.5−Sd
24	28	28−(1.2 Sd)	22.0	22.0−Sd
30	35	35−(1.2 Sd)	27.5	27.5−Sd
36	42	42−(1.2 Sd)	33.0	33.0−Sd

[†] For SI units: one square inch = 645 square millimeters.

*The maximum allowable fill areas in Columns 2 and 4 shall be computed. For example, the maximum allowable fill, in square inches, for a 6-inch (152-mm) wide cable tray in Column 2 shall be: 7 minus (1.2 multiplied by Sd).

**The term Sd in Columns 2 and 4 is equal to the sum of the diameters, in inches, of all 4/0 AWG and larger multiconductor cables in the same cable tray with smaller cables.

(B) Single Conductor Cables in Ladder or Ventilated-Trough Cable Trays
for Cables Rated 2000 V or Less[†]

inside width of cable tray (in.)	*maximum allowable fill area in square inches for single conductor cables in ladder or ventilated-trough cable trays*	
	column 1 (in.^{2})	column 2* (in.^{2})
6	6.50	6.50 – (1.1 Sd)**
12	13.0	13.0 – (1.1 Sd)
18	19.5	19.5 – (1.1 Sd)
24	26.0	26.0 – (1.1 Sd)
30	32.5	32.5 – (1.1 Sd)
36	39.0	39.0 – (1.1 Sd)

[†] For SI units: 1 in.^{2} = 645 mm^2.

*The maximum allowable fill areas in column 2 shall be computed. For example, the maximum allowable fill, in square inches, for a 6-in. (152-mm)-wide cable tray shall be: 6.5 minus (1.1 multiplied by Sd).

**The term Sd in column 2 is equal to the sum of the diameters, in inches, of all 1000 MCM and larger single conductor cables in the same ladder or ventilated-trough cable tray with smaller cables.

Source: Reprinted with permission from NFPA70-1984, *National Electrical Code*®, Copyright © 1983, National Fire Protection Association, Boston, Mass. This reprinted material is not the complete and official position of the NFPA on the referenced subject, which is represented only by the standard in its entirety.

Table 21 Common Electrical Abbreviations

unit	symbol
Alternating current	ac
American wire gage	AWG
Ampere	A or amp
Ballast	BALL
Bill of material	B/M
Circuit	CKT
Circuit breaker	CB
Circular mils	CM
Coefficient of utilization	C.U.
Color code	CC
Conductor	COND
Connection	CONN
Contact	CONT
Control	CONT
Cubic foot	cu ft or ft^3
Current flow	I
Current transformer	CT
Degrees Celsius	$^\circ$C
Degrees Fahrenheit	$^\circ$F
Diameter	Dia or ϕ
Direct current	dc
Disconnect switch	DISC
Double pole	DP
Double pole, double throw	DPDT
Double pole, single throw	DPST
Double throw	DT
Electromotive force	E or EMF
Foot	ft or '
Footcandle	fc
Frequency	FREQ
Ground	GND (GRD)
Hertz	Hz
Horsepower	hp
Hour	hr
Illumination Engineering Society	IES
Impedance	Z
Inch	in. or "
Junction box	JB
Kilovolt-ampere	kVA
Kilowatt	kW
Light	LT
Lumen	lm
Maintenance factor	M.F.
Mega (10^6)	MEG
Megahertz	MHz
Meter	m
Micro (10^{-6})	μ
Milli (10^{-3})	m
Momentary contact	MC
Motor	M or MOT
Motor control center	MCC
National Electrical Code®	NEC
National Electrical Manufacturers Association	NEMA
National Fire Protection Association	NFPA
Negative	NEG

Table 21 *(continued)*

unit	symbol
Neutral	NEUT
Normally closed	NC
Normally open	NO
Ohm	Ω
Overload	OL
Phase	PH
Positive	POS
Power	PWR
Primary	P
Reactive volt-ampere	var
Resistance	R
Secondary	SEC or S
Selector	SEL
Series	SER
Single phase	1 ph
Single pole	SP
Single throw	ST
Supply	SUP
Synchronous	SYN
Terminal	TERM
Terminal board	TB
Thousand circular mils	MCM
Three phase	3 ph
Time delay	TD
Undervoltage	UV
Volt or voltage	V
Volt-ampere	VA
Watt	W

Table 22 Common Electrical Formulas

Ap parent power (single-phase)

$$kVA = \frac{IE}{1000}$$

where kVA = parent power (kilovolt-amperes)
          I   = current (amperes)
          E   = electromotive force (volts)
          1000 = conversion factor (kilo)

Ap parent power (three-phase)

$$kVA = \frac{I \cdot E \cdot 1.73}{1000}$$

where kVA = parent power (kilovolt-amperes)
          I   = current (amperes)
          E   = electromotive force (volts)
          1.73 = $\sqrt{3}$ line voltage conversion factor
          1000 = conversion factor (kilo)

Horsepower (single-phase)

$$hp = \frac{I \cdot E \cdot Eff.}{746}$$

where hp   = horsepower
          I   = current (amperes)
          E   = electromotive force (volts)
          Eff. = efficiency (percent)
          746 = conversion factor (watts/hp)

Horsepower (three-phase)

$$hp = \frac{I \cdot E \cdot Eff. \cdot 1.73}{746}$$

where hp   = horsepower
          I   = current (amperes)
          E   = electromotive force (volts)
          Eff. = efficiency (percent)
          746 = conversion factor (watts/hp)
          1.73 = $\sqrt{3}$ line voltage conversion factor

Illumination

$$fc = \frac{(lm/lamp)\ (no.\ lamps)\ (C.U.)\ (M.F.)}{surface\ area}$$

where fc   = lighting level (foot candles)
          lm/lamp   = lighting output per bulb (lumens)
          no. lamps   = number of bulbs
          C.U.   = coefficient of utilization
          M.F.   = maintenance labor
          surface area = size of the surface (sq. ft.)

Power (resistive)

$$P = IE$$

where P = power (watts)
          I = current (amperes)
          E = electromotive force (volts)

Ohm's Law

$$E = IR$$

where E = electromotive force (volts)
          I = current (amperes)
          R = resistance (ohms)

Table 22 (continued)

Energy	$\text{kWh} = \dfrac{I \cdot E \cdot \text{hr}}{1000} = \dfrac{P \cdot \text{hr}}{1000}$

where

kWh	=	energy (kilowatthours)
I	=	current (amperes)
E	=	electromotive force (volts)
P	=	power (watts)
hr	=	time (hours)
1000	=	conversion factor (kilo)

Table 23   Fractional-Decimal-Metric Equivalents of Length
Conversion Table: Fractions to Decimals

fraction (inch)	decimal 2-place	decimal 3-place	metric (mm)	fraction (inch)	decimal 2-place	decimal 3-place	metric (mm)
1/64	0.02	0.016	0.4	33/64	0.52	0.516	13.1
1/32	0.03	0.031	0.8	17/32	0.53	0.531	13.5
3/64	0.05	0.047	1.2	35/64	0.55	0.547	13.9
1/16	0.06	0.062	1.6	9/16	0.56	0.562	14.3
5/64	0.08	0.078	2.0	37/64	0.58	0.578	14.7
3/32	0.09	0.094	2.4	19/32	0.59	0.594	15.1
7/64	0.11	0.109	2.8	39/64	0.61	0.609	15.5
1/8	0.12	0.125	3.2	5/8	0.62	0.625	15.9
9/64	0.14	0.141	3.6	41/64	0.64	0.641	16.3
5/32	0.16	0.156	4.0	21/32	0.66	0.656	16.7
11/64	0.17	0.172	4.4	43/64	0.67	0.672	17.1
3/16	0.19	0.188	4.8	11/16	0.69	0.688	17.5
13/64	0.20	0.203	5.2	45/64	0.70	0.703	17.9
7/32	0.22	0.219	5.6	23/32	0.72	0.719	18.3
15/64	0.23	0.234	6.0	47/64	0.73	0.734	18.7
1/4	0.25	0.250	6.4	3/4	0.75	0.750	19.1
17/64	0.27	0.266	6.7	49/64	0.77	0.766	19.4
9/32	0.28	0.281	7.1	25/32	0.78	0.781	19.8
19/64	0.30	0.297	7.5	51/64	0.80	0.797	20.2
5/16	0.31	0.312	7.9	13/16	0.81	0.812	20.6
21/64	0.33	0.328	8.3	53/64	0.83	0.828	21.0
11/32	0.34	0.344	8.7	27/32	0.84	0.844	21.4
23/64	0.36	0.359	9.1	55/64	0.86	0.859	21.8
3/8	0.38	0.375	9.5	7/8	0.88	0.875	22.2
25/64	0.39	0.391	9.9	57/64	0.89	0.891	22.6
13/32	0.41	0.406	10.3	29/32	0.91	0.906	23.0
27/64	0.42	0.422	10.7	59/64	0.92	0.922	23.4
7/16	0.44	0.438	11.1	15/16	0.94	0.938	23.8
29/64	0.45	0.453	11.5	61/64	0.95	0.953	24.2
15/32	0.47	0.469	11.9	31/32	0.97	0.969	24.6
31/64	0.48	0.484	12.3	63/64	0.98	0.984	25.0
1/2	0.50	0.500	12.7	1	1.00	1.000	25.4

Index